STORM SURGE

STORM
SURGE

HURRICANE SANDY, OUR CHANGING CLIMATE, AND EXTREME WEATHER OF THE PAST AND FUTURE

Adam Sobel

HARPER WAVE

An Imprint of HarperCollins*Publishers*

HarperCollins books may be purchased for educational, business, or sales promotional use. For information, please e-mail the Special Markets Department at SPsales@harpercollins.com.

FIRST EDITION

Book design by Sunil Manchikanti

Library of Congress Cataloging-in-Publication Data has been applied for.

ISBN: 978-0-06-230476-6

14 15 16 17 18 OV/RRD 10 9 8 7 6 5 4 3 2 1

This book is dedicated to all
those who suffered great losses
as a result of Sandy, and all
those who rushed in to help.

CONTENTS

AUTHOR'S NOTE

I was motivated to write this book after the experience of being interviewed repeatedly by the media in the weeks after the storm. Sandy raised some questions that I could answer from my perspective as an atmospheric scientist. These included questions about the details of the storm itself, and about what current science allows us to say (or does not) about Sandy's relation to human-induced climate change. Other questions were broader, such as: what measures should be taken to protect against future Sandy-like disasters? I realized, by a few weeks after landfall, that I wanted to write about all these questions—both those my technical expertise would allow me to address with some authority and those to which I could give only an opinion as an informed citizen—in a book.

There is a little bit about me in the story. I am your narrator, and you may as well know whom you're dealing with. I've tried not to beat you over the head with my personal politics, but I haven't tried to keep them from showing, either. We scientists, like everyone else, have values, and those are going to come out one way or another. What is important is that we separate normative statements based on those values (statements about what we think people should do) from scientific statements that are not based on those values but that are as objective and evidence-based as we can make them (statements about how nature works). This book contains both kinds of statements. Without being fully explicit or formal about it, I've tried to make it clear enough which are which.

Wind speeds are given mostly (but not exclusively) in knots, following the National Hurricane Center advisories. One knot is 1.15 mile per hour;

65 knots, NHC's estimate of Sandy's winds at landfall (and the minimum threshold for a tropical cyclone to be called a hurricane) is about 75 miles per hour. One knot is also 0.51 meter per second. Distances are measured mostly, but not exclusively, in miles. Storm surge is measured in feet, while sea level rise is measured in both feet and meters.

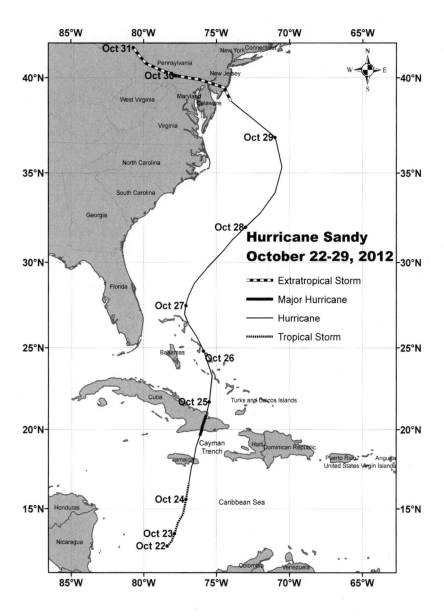

Map showing the track of Hurricane Sandy, starting from the time it was named as a tropical storm, on October 22, 2012. The line type categorizes the storm by type and intensity, as indicated in the legend. *(Courtesy of Andrew Kruczkiewicz and Jerrod Lessel, track data from National Hurricane Center/NOAA)*

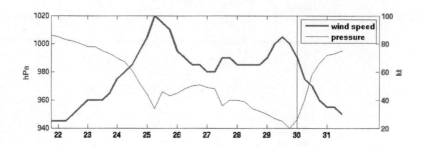

Sandy's maximum wind speed (thick curve, in knots, scale at right) and minimum surface pressure (thin curve, in hectoPascals, scale at left) as functions of time, given as dates in October on the x axis. The time of landfall is shown approximately by the vertical line, indicating 00 Universal Time on October 30. *(Data from National Hurricane Center/NOAA)*

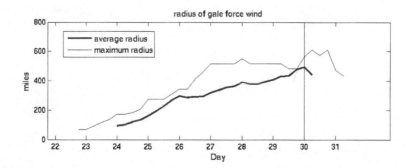

Sandy's size, as measured by the radius of gale force (thirty-five-knot) wind (in miles, on the y axis), as a function of time (date in October, on the x axis). The radius of gale-force wind is measured separately in each quadrant of the storm (northeast, southeast, southwest, northwest); the thick curve is the average of the four values, while the thin curve is the maximum. The time of landfall is shown approximately by the vertical line, indicating 00 Universal Time on October 30. *(Data from National Hurricane Center/NOAA)*

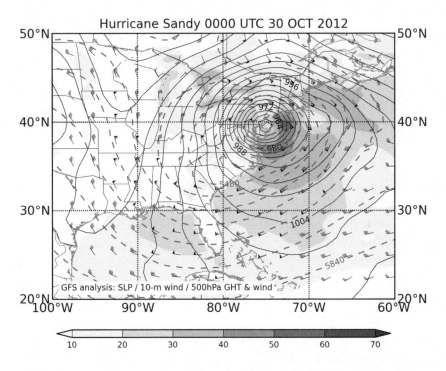

Weather map at the time of Sandy's landfall, 00 Universal Time on October 30, 2012. The solid contours show mean sea level pressure, in hectoPascals; the dashed contours show geopotential height at the 500 hectoPascal pressure level (similar to the pressure field at a constant height of about three miles above the surface); the wind barbs indicate the wind speed in knots (triangular flag = fifty knots, full barb = ten knots, half barb = five knots) at the 500 hectoPascal level; and the shading indicates the wind speed at the surface, in knots, with the scale given at bottom. (© 2014 Chia-Ying Lee, data from NOAA's Global Forecast System)

INTRODUCTION

On October 28, 2012, a giant, misshapen hurricane made a left turn from its previous northward trajectory over the Atlantic Ocean and headed for the New Jersey coast. On the evening of October 29, following a track never before seen in one hundred sixty years of Atlantic hurricane observations, the center of the storm made landfall near Atlantic City. The destruction sprawled far from there.

A few days earlier, over the Bahamas, Sandy had crossed paths with another weather disturbance, a low-pressure system in the upper atmosphere. It took on some of that system's properties, including its great size. As it made landfall, it merged with yet another system, an extratropical, or "winter," storm that had come from the North American continent at the leading edge of a blast of autumn cold air. This merger gave Sandy a new jolt of energy, increased its size yet further, and completed its transition from a tropical cyclone to a mammoth hybrid. When reporters saw this coming in the weather forecasts, they dubbed it, glibly at first, Frankenstorm. But as its gravity soon defied humor, it was renamed Superstorm Sandy.

The size of the storm, like the track, was unprecedented in scientific memory. Sandy was the largest hurricane ever observed in the several decades since good measurements of hurricane size have existed in the Atlantic. At its landfall, gale-force winds covered a large fraction of the Eastern Seaboard and an enormous patch of oceanic real estate as well. To the north of the center, Sandy's easterlies traversed a thousand-mile-plus fetch before coming onshore, driving a massive storm surge: a giant, slow wave that dragged the ocean inland, on top of the high tide, and onto some of the most heavily populated, economically active, and valuable land on earth.

The scale of the disaster was historic. In New York City, the water had not come this high since at least 1821, if then. For people in the hardest-hit areas, it was a life-crushing event—in some cases, literally. While the death toll was low compared to Hurricane Katrina's, and extremely low compared to those of the worst tropical cyclone disasters in recent history worldwide, it was high enough to be grievously shocking here in New York City, where losing one's life to a hurricane is thought of as something that happens only in faraway places. Many, many people saw their homes destroyed, and in some cases entire neighborhoods. The storm crippled the infrastructure of one of the world's most vibrant economic and cultural centers for a period of weeks. The economic damage has been counted at fifty billion dollars at least, and perhaps as high as sixty-five billion.

Was Sandy a freak of nature or a sign of disasters to come? Where did it come from, and what were the atmospheric forces that made it happen? How did we manage to live for so long in such a vulnerable state, knowing as we did that storms like this were possible? Now that we know, not just theoretically but also from hard experience, that such storms can happen here in New York City (and in other coastal cities that have never had weather disasters of this magnitude, but well could), what should we do to prepare for the future? This book is an attempt to answer these questions by telling the story of the storm, and putting it in both scientific and historical context.

The most fundamental lessons we can draw from Sandy revolve around predictions: how we make predictions of the atmosphere's behavior, and how we respond to them once they are made. Weather prediction is a unique enterprise. People make predictions of many kinds: about the outcomes of elections or baseball games, or the fluctuations of the stock market or of the broader economy. Some of those forecasts are based on mathematical models. Most of those mathematical models are statistical, meaning they use empirical rules based on what has happened in the past. The models used for weather prediction (and its close relative, climate prediction), in contrast, are dynamical. They use the laws of physics to predict how the weather will change from one moment to the next. The underlying laws governing elections or the stock market—the rules of mass human behavior that determine the outcomes—are not known well, if they exist at all. The models need to be built on assumptions that past experience will be a guide to future performance. If weather prediction were still done in this way, it would have been simply impossible to predict, days ahead of time, that Hurricane Sandy would turn left and strike the coast while moving westward. No forecaster had ever seen something like that occur, because no storm had ever done it. For the same reason, no statistical model trained on past behavior would have produced it as a likely outcome.

In Sandy's case, forecasters not only could see this outcome as a possibility over a week ahead of time, but they were quite confident of it by four or five days before the storm hit. Forecasts such as the ones we had as Sandy formed and moved up the coast don't come from the heavens. They're the result of a century of remarkable scientific achievement, beginning in Norway in the early 1900s. The intellectual foundation of the whole enterprise of weather prediction was the idea that the laws of physics could be used to understand the weather, a radical idea in the early twentieth century. Carrying this out required multiple conceptual advances, over decades, and improvements in technology (especially digital computers).

As good as weather forecasts have become, we are still learning how to use them right. Chaos theory, developed by Ed Lorenz in the 1960s, tells us that even though our weather models are based on exact physical laws, the forecasts must nonetheless contain some uncertainty, no matter how good the forecasters are. How best to communicate uncertain forecasts, and how those in positions of responsibility should best make decisions in the face of that uncertainty, are questions on which Sandy shone a bright light. The light gets brighter and more uncomfortable when we look farther ahead into the future than the ten-day weather forecast.

The measures taken by local, state, and federal government agencies and officials to prepare for Sandy in the days before landfall, and those taken to respond in the days and weeks after it happened, were very successful by many measures. The death toll in the United States (117 according to the Centers for Disease Control), awful as it was, could easily have been much, much higher. The immediate aftermath, hard as it was for many (and still is for some, as I write a year and a half later), was orderly on the large scale. There were no episodes of mass looting or other crimes, no large-scale food or water shortages, no epidemics of disease. Compared to Katrina in 2005 (not to mention many relatively recent tropical cyclone disasters in the developing world), Sandy was handled incredibly well, in the short term.

Not all went right, of course. Many people didn't evacuate who should have. Some of them died, either in the storm itself or in the disruptions that followed. The power outages, the flooded transportation tunnels, the gas lines, and the elderly and vulnerable people left stranded on the upper floors of dark nursing homes and public housing projects—all these showed us that we don't quite have disaster management down to a science.

But the most serious problems highlighted by Sandy were not in the preparations right before the disaster or in the response right after. They were in the construction of our coastlines over the span of many decades. Over that long term, too, there had been good forecasts of what could happen to our built environment along the water in the New York City area. These were not forecasts of a specific storm at a specific date and time, but rather scientific assessments of the *risk* of a storm as bad as Sandy, or worse. It had been known for decades at least that New York City was vulnerable to flooding by a hurricane-induced storm surge. The consequences that would follow were also clear, in broad outline. The flooding of the subways, for example, had been envisioned since the 1990s.

But apparently we don't take these kinds of long-term risk assessments nearly as seriously (if seriousness is measured by action) as we take forecasts of an impending storm. This problem appears in yet more acute form when we try to deal with human-induced climate change.

Sandy didn't need climate change in order to happen, and the story of the disaster doesn't need climate change to make it important. The main subject of this book is Sandy, and you can read large fractions of the book without seeing climate change mentioned at all. But climate change looms large when we try to think about what Sandy means for the future.

Our first task, though, is to understand what actually happened in the atmosphere to produce Sandy.

The global atmosphere is always in a complicated state of motion: jet streams and trade winds rippling and swirling with giant eddies and waves. Much of that complex motion is chaotic and hard to put a label on, but a few coherent patterns repeat themselves often enough that we give them names. Some, such as El Niño, are household words. Others aren't but should be.

Two of the most important ones, the Madden-Julian Oscillation (MJO) and the North Atlantic Oscillation (NAO), were active in the days before Sandy formed in the Caribbean, and they set the stage for the storm's genesis. We'll see how these natural climate fluctuations made Sandy not just an Atlantic hurricane, but a complex atmospheric event with global dimensions. The MJO pattern that drove Sandy's formation had traveled eastward into the Caribbean from the tropical western Pacific, while the strong NAO had pushed the jet stream south from the polar regions to create the conditions for Sandy's rejuvenation late in life as a post-tropical hybrid "superstorm."

After Sandy formed and began to make its way north, it began to feel the presence of a disturbance in the autumn jet stream. This was the first extratropical, or winter, storm over the United States that would influence Sandy, still down in the Bahamas. Like someone in a long-term relationship whose personality becomes more like her partner's over time, Sandy began changing its character to acquire some of the traits of the winter storm. It would change completely when it met its second partner, just before landfall.

Sandy was a departure from all known meteorological history not just because of how it changed into a hybrid—other storms have done that—but because of the left turn it took as it did so. Atlantic hurricanes that merge with winter storms usually make their way harmlessly out to sea. Sandy's violation of that pattern was what made it truly fair to call it a freak. This radical track was a result of interactions with both the blocking high-pressure system offshore and the winter storm coming from the continent. As the blocking high was cutting off Sandy's escape to the east, the winter storm was developing to the west. The two storms drew closer, until they got into each other's orbits, literally. That orbiting process, known as the Fujiwhara effect, occurs when two nearby vortices circle around each other. In the case of Sandy, it was a dance that resulted in something we've never seen before, and that defied the standard characterizations of our storm warning system.

But what does Sandy portend for the future?

If Sandy was a freak storm, so rare that it was like none we had seen before, does that mean that the next one isn't likely to occur until the far future, or are we in for many more of these now? Sandy hit just one year after another hurricane, Irene, made landfall in New York City. By coming in more weakly than forecast, Irene just barely let the city off the hook; but its tremendous rains fed inland floods in the mountainous areas of northern New England and upstate New York, wreaking destruction that was as terrible and as unprecedented for those inland regions as Sandy was for New York City and New Jersey. Are frequent, destructive storms in the northeastern United States (or in other places that aren't accustomed to them) the "new normal" as the climate warms and the coasts get built up? This question breaks down into several different questions that we need to tackle one at a time.

The first question is whether there will be more hurricanes, or stronger ones, in the warmer climate that is coming. This turns out to be a very difficult science problem. To answer it, we need to understand the relationship between hurricanes and the larger climate in which they occur. Until around ten years ago, the study of the relationship between climate and hurricanes was, perhaps surprisingly, a fairly undeveloped backwater of atmospheric science. Then it blew up, especially after Katrina. The research effort expended on it since then has dwarfed what came before, and some real progress has resulted.

If what made Sandy so freaky was the set of circumstances that led to that left turn, then what about that? Is that trajectory something we can expect to happen more often in a greenhouse world? Will blocking highs and mergers with winter storms send more superstorms careening toward the northeastern United States (or other coastal areas), even if there aren't more storms altogether? Recent research has just begun to delve into this question.

There is one connection between Sandy and our warming climate that is simple and undeniable: the rising sea. As our planet gets warmer—and it is getting warmer, due to human emissions of greenhouse gases, no doubt about it—the ocean is swelling up in the heat. It's a tiny effect on the face of it—a degree of warming increases the volume of seawater by a small fraction of a part in a thousand—but it turns out to be enough to matter: it has raised the sea level by inches already, and will continue to do so as the oceans warm further. At the same time, ice on land is starting to melt, increasing the amount of water in the ocean. The water has been rising inexorably for a century, and it's going to keep rising. The surge from future storms will pile on top of that new higher water level and give us worse floods, even if the storms themselves don't change.

The only question is how fast the sea will rise. It could be slow, if we're lucky. Or big chunks of the earth's giant polar ice sheets, on Greenland and Antarctica, could start to break up and fall into the ocean faster than we used to think possible. This is the truly scary scenario that our polar scientists have only recently begun to grapple with seriously. In this book, we'll try to understand what will determine how much time we have until the water washes over the seawalls in Lower Manhattan, even on clear days.

Finally, we'll look at some of the engineering solutions that have been put in place in the last century to protect vulnerable coastlines against storm surge, and the historical events that led to them. The most ambitious examples are in the Netherlands, where multiple storm surge barriers, massive and sophisticated, have been built. But here in the northeastern United States, we, too, have a few of our own, smaller barriers, as well as artificial dunes and other less dramatic measures. Huge, Dutch-style barriers have been advocated for New York Harbor now; so have many other, less "hard" solutions, including "retreat" (moving people out of the most vulnerable areas).

History shows quite clearly that human beings are not good at reacting with great foresight to long-term risks of low-probability, high-impact events until these have happened at least once or twice, even if the risk is known and understood earlier. The abstract scientific knowledge that a disaster will occur, sooner or later, seems to be inadequate to motivate investment in protective measures, no matter how clearly it can be shown that these measures will pay for themselves in the long term. Most of us just can't visualize the disaster until we actually see it. Lacking that visualization, governments view protection from disaster as a low priority. Only in the brief historical moment after a disaster happens does long-term thinking have a chance.

What will Sandy's legacy be? What new protections will be put in place, now that imagination is no longer needed to visualize the New York metropolitan area underwater?

In any serious discussion of new protective measures, the reality of global warming can't be avoided. Sea level rise, especially, changes the calculus. This makes the discussion now different from any that took place after earlier storms, and it makes our inability to visualize disasters before they happen an even worse handicap now than in the past.

Global warming, just like a single storm worse than any in the last one hundred years, will bring impacts outside the experience of anyone alive. Unlike a single storm, though, global warming happens gradually; and unlike a single storm, global warming has elicited a powerful, well-financed, and (thus far) partly successful disinformation campaign whose goal is to deny that it is occurring and to prevent any action from being taken in response to it. It seems possible that future generations may see Sandy's landfall as a significant historical landmark in the evolution of the public perception and government response to climate change in the United States—or not.

We're now in a fascinating, if somewhat frightening moment. Complex questions involving science, engineering, politics, and human psychology are swirling around Sandy figuratively, like its winds did literally. I can't answer all those questions, but this book is an attempt, one scientist's attempt, to grapple with some of them.

PART ONE
THE STORM

1

GENESIS

The Disturbance

A LARGE AREA OF DISTURBED WEATHER OVER THE EAST-
ERN AND CENTRAL CARIBBEAN SEA IN ASSOCIATION WITH A
WESTWARD-MOVING TROPICAL WAVE. ENVIRONMENTAL CON-
DITIONS ARE FAVORABLE FOR SLOW DEVELOPMENT OF THIS DIS-
TURBANCE DURING THE NEXT FEW DAYS. THIS SYSTEM HAS A
LOW CHANCE . . . 10 PERCENT . . . OF BECOMING A TROPICAL CY-
CLONE DURING THE NEXT 48 HOURS.

Forecaster Todd Kimberlain wrote this in the National Hurricane Center's Tropical Weather Outlook issued on Friday, October 19, 2012, at 2:00
p.m. Eastern Daylight Time.

The words were accompanied by a grayscale infrared satellite image
showing the clouds over the North Atlantic Ocean, with the Americas on
the left side and a little slice of West Africa on the right. Superimposed on
the image was a blue box enclosing text in a bold white font: "Tropical Cyclone Activity Is Not Expected During the Next 48 Hours." Kimberlain had
judged that a storm was gradually getting itself together, but also that it was
outside the forty-eight-hour time horizon of the Tropical Weather Outlook,
a standard NHC product issued four times daily.

The situation had changed by the next Outlook, issued at 8:00 p.m. and
written by forecaster Stacy Stewart. That one began:

FOR THE NORTH ATLANTIC . . . CARIBBEAN SEA AND THE GULF
OF MEXICO . . .

1. A WESTWARD-MOVING TROPICAL WAVE LOCATED SOUTH OF
HISPANIOLA IS INTERACTING WITH AN ELONGATED TROUGH OF
LOW PRESSURE THAT EXTENDS FROM NICARAGUA EASTWARD
TO THE SOUTHERN WINDWARD ISLANDS. THIS BROAD DISTUR-
BANCE IS PRODUCING WIDESPREAD CLOUDINESS AND SHOW-
ERS OVER MUCH OF THE SOUTHWESTERN . . . CENTRAL . . . AND
EASTERN CARIBBEAN SEA . . . AND ADJACENT LAND AREAS OF
COLOMBIA AND VENEZUELA. OVER THE NEXT FEW DAYS . . . EN-
VIRONMENTAL CONDITIONS ARE EXPECTED TO GRADUALLY
BECOME MORE CONDUCIVE FOR DEVELOPMENT TO OCCUR
ACROSS THE CENTRAL AND SOUTHWESTERN CARIBBEAN SEA.
THIS SYSTEM HAS A LOW CHANCE . . . 20 PERCENT . . . OF BECOM-
ING A TROPICAL CYCLONE DURING THE NEXT 48 HOURS AS IT
MOVES SLOWLY WESTWARD.

The blue box containing the statement that tropical cyclone activity was
not expected had been deleted. Instead, a green oval had been drawn around
the large cloudy region in the Caribbean, south of Cuba and Hispaniola and
north of South America. The oval delineated where a tropical cyclone had a
20 percent chance of forming in the next forty-eight hours. This image, the
Graphical Weather Outlook, is shown in the color insert.

The "tropical wave" was a typical weather system for the Caribbean at
this time of year. Also known as easterly waves, such systems disturb the
otherwise steady trade winds that blow from east to west (i.e., easterly) at
low latitudes.

Easterly waves often form over Africa, south of the Sahara but north of the
equator. There, the earth's normal temperature gradient reverses. Almost everywhere on earth, the climate cools as one moves away from the equator and
toward either pole; in western Africa, the contrast between the cooler, wet
jungle to the south and the hot desert to the north breaks that rule. Coupled
to that reversed temperature contrast is a strong "jet," or narrow, strong
air current, blowing easterly (from east to west) and centered a few miles
aboveground. The African easterly jet is congenitally unstable; it can't blow

steadily in a straight line. It spontaneously develops undulations hundreds or even thousands of miles across. These are easterly waves. Blown from east to west by the jet, these waves bring intense squall lines, dust storms, lightning, and thunder; they also bring welcome rain to the nations of the semiarid Sahel, on the boundary between the Sahara and the wet Guinea coast.

Then the easterly waves leave Africa for the tropical Atlantic Ocean. The jet weakens as it escapes the continent. The waves can weaken as well, but some hold together as they drift toward the Americas. The cold tops of deep cumulus clouds, indicative of heavy tropical rain beneath, are sometimes apparent in satellite images, sometimes not—sometimes there is nothing but the circulation, not directly visible from space, a gentle curve in the trades as they traverse the area, at least several hundred miles wide, of very slightly depressed surface pressure.

Sometimes the cumulus clouds begin to flare more strongly and persistently. The curve in the winds becomes less gentle and wraps around to become a closed circulation, the clouds begin to coalesce and deepen around the circulation's center, the pressure at the center drops, and the system organizes and strengthens itself to the point that we call it a tropical storm, and then a hurricane.

Sometimes that happens; more often it doesn't. Stewart's judgment in this case was that "over the next few days . . . the environmental conditions" were "expected to become gradually more conducive for development." In other words, he expected that the odds of a hurricane's forming out of the tropical wave were going to increase, because of some "environmental conditions." The Outlook didn't specify what those conditions were or why they were expected to change in a way that would favor the birth of a hurricane.

The Environment

Though NHC didn't issue forecasts of tropical cyclone formation further than two days out in 2012, colleagues at the Climate Prediction Center, in College Park, Maryland, had been issuing longer-range forecasts since around five years earlier. The CPC Global Tropics Hazards and Benefits Outlook uses data from many sources to provide projections of weather throughout the tropics, with a time scale of two weeks. These Outlooks don't make specific predictions for any particular day, but only probabilistic statements about a whole week at a time.

For week one, the first week after the day the Outlook is issued, and for the next one, week two, the Outlook rates the confidence with which either above-normal rainfall or below-normal rainfall is predicted as "high" or "moderate" in specific areas. The areas of these forecast anomalies are marked by blobs of color on a map: two maps, one for each week. Areas where above- or below-normal temperature is also highly to moderately probable are similarly marked, in different colors. The color red is reserved for areas where "tropical cyclogenesis" (the birth of a tropical cyclone) is projected to be likely. Candy cane red-and-white stripes indicate moderate confidence of tropical cyclone formation; solid red, high confidence.

In the Outlook issued on October 9, 2012 (ten days before Stewart's and Kimberlain's Outlooks identifying the easterly wave with the potential to become Hurricane Sandy), tropical cyclone formation was marked for week one (October 10–16) with a "moderate confidence" red-and-white oval in an eastern Pacific region offshore southern Mexico. Another, smaller region of moderate confidence was drawn in the western Atlantic, just east of the Antilles. In the week two map, the Atlantic blob was gone.

Just as in the shorter-range NHC Outlooks, the maps in the CPC Outlooks are followed by a discussion. The first paragraph of the CPC Outlook of October 9, written by forecaster Jon Gottschalk, began:

> The latest observations indicate that the MJO remains weak. Although considerable enhanced convection continues across the western Pacific, there has been little evidence to date of this convection shifting eastward coherently on the MJO time scale, as indicated by a few different measures. The majority of dynamical model MJO index forecasts indicate a continuation of a weak signal during Week-1 with some agreement for a stronger signal in the index to emerge during Week-2 across the western Hemisphere.

And continued:

> Model forecasts of the MJO index have been relatively poor after Week-1 in recent weeks. Based on this the MJO is forecast to remain generally weak through most of the outlook period and at this time not expected to contribute substantially to anomalous tropical convection during the next 1–2 weeks.

The Outlook continued for three more paragraphs, the last of which described the tropical cyclone formation probabilities shown on the map for the Atlantic and eastern Pacific.

The discussion in the following week's CPC Outlook, issued on October 16 by forecaster Brad Pugh, showed a change in this thing called the MJO. It began:

> The latest observations indicate that the MJO remained weak during the past week. However, some observations recently indicate a more coherent MJO signal. A large spread exists among dynamical model MJO index forecasts. Some forecasts favor little coherent signal, while other forecasts indicate an increase in amplitude of the MJO index and eastward propagation. Due to the large model spread and poor performance of model forecasts during October, uncertainty is high during the next two weeks.

The end of the third and beginning of the fourth and last paragraph read:

> Late in the week-1 period, the chances for tropical cyclone development are forecast to increase across the western Caribbean.
>
> MJO composites, warmer-than-normal SSTs, and climatology favor tropical cyclone development in the western Caribbean during week-2. The GFS model has been very consistent with tropical cyclone development in the western Caribbean early in week-2.

The week one map did not explicitly show any area of tropical cyclone formation, perhaps because the likelihood was not expected to increase until late in the week. But the week two map showed a solid red blob, angled northwest to southeast, filling the chunk of ocean between Central America and Cuba.

"GFS model" refers to the numerical weather prediction model known as the Global Forecast System. The GFS was the primary U.S. weather forecast model, run by another National Oceanic and Atmospheric Administration (NOAA) lab, the Environmental Modeling Center, part of the National Centers for Environmental Prediction, just next door in College Park. The GFS was run every six hours, starting from the latest atmospheric conditions. Each run simulated the next few weeks of the atmosphere's behavior.

Pugh's statement that the GFS model had been "consistent" indicated that although the tropical cyclone development in the model was more than a week in the future (and thus, uncertain), it had been happening in run after run. That still didn't necessarily mean the model was right, but it did mean the model was making a strong statement.

This CPC Outlook issued on October 16 and the NHC Outlook issued at 8:00 p.m. EDT on October 19 were in complete agreement. The CPC had said in its weekly Outlook that the chances for cyclone development would increase late in week one. NHC's forty-eight-hour Outlook issued on the evening of the nineteenth, late in the CPC's week one, said that the environment was becoming more favorable for cyclone development. The forecasters in the two NOAA laboratories were looking at the same information.

Apparently, the most important piece of information was the MJO. Both the October 9 and October 16 CPC Outlooks began with the status of this thing. They didn't explain what the MJO is, or even spell out the words to which the initials refer. But the MJO was the most significant factor increasing the likelihood of a tropical cyclone. The CPC also cited "warmer-than-normal SSTs," referring to sea surface temperatures, and to "climatology," meaning what usually happens at that time of year. But neither the SST nor the climatology can change much in a few days. Those factors couldn't explain the change in the environment from one week to the next. The MJO was making that change. The MJO was the influence, moving into place, that would nurture our late-season Caribbean hurricane into existence.

2

MJO

In the late 1960s, Roland Madden and Paul Julian were listening for the bass line in the tropical atmosphere.

Over two decades earlier, World War II had stimulated U.S. government interest in meteorology over the tropical oceans. The military had fought hard campaigns in the near-equatorial Pacific, during which it needed to contend with the weather there. This weather was much different from anything in Europe or the United States. Little was known or understood about it.

During the war, meteorology was still in the first adolescent decades of its growth into a modern science. The new insights driving the field forward had come mostly from Scandinavia. Concepts such as the polar front, important though they were in northern Europe, were useless in the tropics. The military's need for someone who could make a forecast was urgent, and the knowledge vacuum almost total.

Clarence Palmer, a zoologist from New Zealand,[1] had only a modest amount of meteorological training before he was drafted into the Royal New Zealand Air Force in 1939. Soon after that, his reputation grew to the point that he was hired to train U.S. Army forecasters at the University of Chicago's meteorology department. Palmer was then made director of the Institute of Tropical Meteorology (ITM), established in 1943 in Rio Piedras, Puerto Rico, by the University of Chicago and the University of Puerto Rico. The department in Chicago was run by Carl-Gustav Rossby, a Swede now recognized as one of the founding fathers of modern meteorology. Rossby had convinced the army to set up the ITM. One of the world's top meteorologists hired someone with almost no formal training to head what immediately was the leading institute in its field; it was the scientific Wild West.

The instruction program was put together by Palmer and Gordon Dunn, a forecaster from the Weather Bureau (the previous name for the National Weather Service). Dunn had been on duty during the 1938 hurricane that devastated Long Island and New England, was the first to clearly document Atlantic easterly waves in the scientific literature, and would become the first director of the National Hurricane Center when it was established in 1965.

The instruction program was two months long, first given in March–May 1943. One of the students, Reid Bryson, commented later that by the end of the two months, the instructors had taught the students everything they knew.[2]

Also among the students was Herbert Riehl. Riehl was a German immigrant who had just completed a degree in meteorology at New York University. He would go on to become the leading tropical meteorologist of the postwar period. He first taught at the University of Chicago, and then moved to Colorado State University in 1961, founding the Department of Atmospheric Science there in 1962. Riehl was the first to conclude that hurricanes get their energy from the ocean. Among his students would be Joanne Malkus (later Joanne Simpson), the first woman in the United States to obtain a PhD in meteorology, whose work with Riehl showed for the first time the critical role of cumulus clouds in the tropical circulation and who would later pioneer the satellite measurement of tropical rainfall; Robert Simpson, later to become director of the National Hurricane Research Project, predecessor to NHC, in 1956; and William Gray, pioneer of seasonal hurricane forecasts (predictions not of a single storm, but of the overall level of activity in an entire season, months in advance), who would himself train several generations of hurricane scientists and forecasters.

After the war ended, many new weather-observing stations were established on Pacific islands. These were tiny specks of land with exotic names: Kwajalein, Truk, Eniwetok, Pohnpei. Some had been the sites of gruesome battles between Allied and Japanese forces. Observers at these stations began launching weather balloons, or radiosondes,[3] to take measurements in the upper atmosphere once or twice per day. By the 1960s, for the first time, there were records multiple years long: consistent, high-quality measurements of temperature, pressure, wind, and humidity, up to ten miles or more above the surface, at many locations in the tropics.

Better tropical weather maps than any before could now be drawn, including maps of the upper-air circulation patterns. More valuable still, the long continuous records allowed one to study the evolution of the weather from one day to the next. Besides looking for patterns in space, it was now possible to look for patterns in time.

Madden and Julian were members of the second generation of modern tropical meteorologists, working in the late 1960s as researchers at the National Center for Atmospheric Research, (NCAR), in Boulder, Colorado. (The main building in which NCAR sits, up on a mesa above town, was made famous by Woody Allen in *Sleeper*; it's the futuristic palace from which his character, Miles Monroe, tries to kidnap the nose of the deceased evil leader of the brainwashed society into which Miles has time-traveled from the 1970s. Maybe the movie made too big an impression on me when I was young, but NCAR's Mesa Lab still looks futuristic to me today.)

While working full-time at NCAR in Boulder, Madden did his PhD at Colorado State, in nearby Fort Collins. Riehl taught the first course he took, in 1968.[4] Much earlier, Julian, growing up in La Porte, Indiana, in the 1940s, had a high school music teacher whose son had been in the military and had attended the training course in Rio Piedras. The written materials from the course had made it home to La Porte, including a short section by Palmer on the tropics. The teacher knew of Julian's interest in weather and gave him the papers in 1946. Julian studied them, fascinated, and his determination to pursue meteorology as a profession was solidified.[5]

Both Madden and Julian had been trained in modern methods of statistical analysis. Julian, in particular, had done his PhD at Pennsylvania State with Hans Panofsky, a specialist in turbulence. Studies in turbulence at the time were (and still are) heavily dependent on statistical methods. These included spectral methods, a set of techniques for finding the oscillating patterns in data. Julian had become expert in these.

Everything that varies in time can be broken down into components with well-defined frequencies. A single, pure musical tone is a sound wave that makes the air vibrate at a single frequency. The A on the musical scale, which is used as the most standard reference for tuning pianos and other instruments, has a frequency of 440 Hertz. That's how many times per second it makes your eardrum move back and forth.

Any real musical instrument produces not a pure single frequency but a combination of that frequency and its harmonics, which are multiples of that frequency. For example, 880 Hertz, twice 440, is the A an octave up; other, less regular multiples produce different notes on the musical scale. A single keystroke on the piano or a single twang of a guitar string produces a combination of frequencies, each with a different amplitude, or strength. The fundamental note (the lowest one sounding) has the greatest amplitude. It determines which note you hear. The octave up (with twice the frequency of the fundamental) will have a lesser amplitude, and other harmonics, all with frequencies higher than the fundamental, will have lesser amplitudes still. You don't hear the harmonics distinctly, but they give the note its distinct sound, or tone color, which is different for each musical instrument. If a few notes are played together, more different frequencies sound. When the members of a whole orchestra (or rock band) play together, you get a very complex sound signal.

Even most music, though, still has only a distinct set of frequencies. There are many notes that are not heard. On the other hand, if one were to have every member of a large orchestra play a different note at random, with some of them choosing notes not even on the traditional scales (i.e., playing "out of tune"), most of us would call the result noise. In precise terms, noise is sound whose amplitude is distributed evenly over a wide range of frequencies. You can't make out individual notes, because all of them are sounding at once.

Even if we don't hear distinct notes when we listen to noise, we might still perceive a general pitch range. There are high-pitched and low-pitched noises. This means that we are hearing not specific frequencies, but more power in broad ranges of frequencies, or bands, than others. If you use an equalizer to boost the treble or bass or midrange on some music you are listening to, you are adding amplitude not to a specific frequency but to a broad range of frequencies.

Any data can be treated as if they were sound. The methods of spectral analysis make it possible to take any sequence of numbers occurring in time and break it down into frequencies. One can determine the amplitude, or power, at each frequency. This breakdown is called the power spectrum.

If the sequence of numbers represents something that is truly periodic, the power spectrum will show a sharp spike at a particular frequency, indicating that a large fraction of the power in the data is at that single frequency. If one were to turn the time series into sound, one would hear a single musical note.

If we were to compute, for example, the power spectrum of the time series of solar radiation at the ground somewhere, it would have a spike at a frequency of one cycle per day.[6] Though clouds complicate things a little, the diurnal cycle is the dominant factor controlling the amount of solar radiation: day is light, night is dark, extremely reliably, repeating every twenty-four hours. If one were to turn that time series into sound, that frequency (one cycle every 86,400 seconds) would be much too low for our ears to hear. But if we could hear it, it would sound like a clear single note. We could use it to tune a very low-pitched orchestra.

Madden and Julian were taking advantage of the new tropical radiosonde data to look at the spectral properties of the weather over the tropical oceans. Did the atmosphere have any coherent variations at particular frequencies? It was an exploratory analysis. There was no reason to expect strong peaks at any particular frequencies other the diurnal cycle (day/night) and the annual cycle (the seasons). Apart from those, one might reasonably expect the winds, for example, to have a power spectrum that looked like noise. This is the case in normal turbulent fluid flows. Many phenomena in atmosphere, ocean, and earth science, too, have power spectra close to red noise—that is, noise with more power at low frequencies than high, like a gigantic orchestra playing every possible note at the same time, but with more tubas and double basses than trumpets and violins.

If one were to find spectral peaks on top of the red noise (frequencies or narrow ranges of frequencies where there is significantly more power than other frequencies nearby), this would suggest the existence of weather patterns that repeated themselves with some regularity. A broad peak would mean that the repetition was not exactly periodic, so that one couldn't predict the next cycle just by watching the clock, as one could with the diurnal and annual cycles.

If one were to find a broad spectral peak in wind data at frequencies in the neighborhood of once every week, this would indicate that the wind blew in one direction for a few days, then usually switched to the other direction. If one finds similar spectral peaks at many stations on different islands, one could then check if the variations between them are coherent. Do the wind shifts happen at the same time at all the stations, or sequentially at one, then the next one nearby, then the next? Or do they all happen randomly at unrelated times?

If there is some coherence, more statistical techniques can be used to determine what variations in flow patterns—sequences of weather maps—are typically associated with the oscillations. One can then hope to use the laws of physics to understand what makes those oscillating weather patterns tick.

In 1966, Taroh Matsuno, a theoretical meteorologist at the University of Tokyo, published a paper describing flow patterns for four types of weather disturbances that, according to his elegant analysis of the equations of motion, ought to occur near the equator. Matsuno's analysis treated the atmosphere as though it were a thin layer of liquid. In his paper published the following year, 1967, Richard Lindzen showed that similar disturbances should occur in a more realistic gaseous atmosphere. It was not immediately clear that these theoretical disturbances, or waves, corresponded to any real weather patterns known to forecasters. But the best observational meteorological researchers of the time immediately recognized the significance of Matsuno's and Lindzen's work.

The new long-term tropical radiosonde data sets made it possible to look for the patterns that Matsuno and Lindzen predicted. Michio Yanai, Matsuno's colleague in Tokyo, and John M. ("Mike") Wallace, at the University of Washington in Seattle, did that. Each type of wave had not only a specific pattern in the maps of winds and pressure, but also a specific relationship between its frequency and its wavelength. Both Yanai and Wallace, in several papers published in the few years after 1966, found evidence for some of the theoretical waves in the data from the islands. The evidence was clearest high up in the stratosphere, above the clouds.

In the troposphere (the lowest ten miles or so), Yanai and Wallace found coherent patterns that seemed to correspond to weather disturbances known to forecasters as easterly waves. It was not clear whether these were the same as Matsuno's or Lindzen's theoretical waves. They were similar to the easterly waves that form in the African easterly jet and move westward over the Atlantic, forming the seeds for Caribbean-born hurricanes such as Sandy. Besides moving from east to west, both types repeat every few days or so. These are high notes in the tropical atmospheric spectrum.

Roland Madden began doing his own spectral analysis of radiosonde data during this time. He had first obtained forty-seven days' worth of data from a field campaign, the Line Islands Experiment, carried out from March to April 1967 in the eastern Pacific, a thousand miles south of Hawaii. He presented his results at a conference in Hawaii, the Symposium on Tropical Meteorology, in June 1970. Even then, when computers were much, much less powerful than those today, forty-seven days' worth of data was not a lot. Madden was told by an older expert in the audience, sternly, that he should find more.

Madden went back to Boulder and got ten years of radiosonde data (1957–1967) from Canton Island, in the central equatorial Pacific (3 S, 172 W). He and Julian carried out a sophisticated spectral analysis using a new mathematical algorithm, the fast Fourier transform (FFT), developed just a few years earlier. (The FFT, invented by electrical engineers, would have a revolutionary impact on virtually every field of science and engineering in the years to come. It drastically reduced the amount of computation required to do spectral analysis, a method with applications to nearly everything.) Madden and Julian used the FFT, the long data set, and the computer at NCAR to look for lower frequencies than others had in the tropics (the bass notes). Instead of looking just for weather patterns repeating every few days, they could look for patterns with frequencies of months. They published their results in the *Journal of the Atmospheric Sciences* in July 1971. I had just turned four years old.

Madden and Julian found something that neither Yanai nor Wallace had. Nor had it been predicted by Matsuno or Lindzen. They found a strong, clear spectral peak, a frequency range with power well above those nearby, at the low-frequency range corresponding to periods in the range of forty to fifty days. In other words, their analysis showed that at Canton, the weather tended to shift systematically back and forth with some regularity, repeating itself somewhere in between once a month and once every two months. We now call this period intraseasonal because it's shorter than a season. There was not, and still is not, any obvious reason to expect anything in the atmosphere to repeat itself with that forty- to fifty-day periodicity. It has no obvious relationship to anything about the orbit or rotation of the earth, or the moon, or any other external influence on the atmosphere.

The oscillations were coherent in multiple meteorological variables, including surface pressure and the zonal (east–west) wind at both low and high levels in the atmosphere. Some aspects of the oscillation bore some resemblance to the equatorial Kelvin wave, one of the theoretical types predicted by Matsuno and Lindzen. Others did not. Madden and Julian concluded that it was not a Kelvin wave, but something else not predicted by any theorist.

Because they had analyzed data from only a single island, Madden and Julian could not tell anything directly about the spatial structure of the oscillation. What kind of weather maps would it produce? Had the same oscillation they had found at Canton occurred elsewhere? If so, did it oscillate in phase everywhere at the same time? Or was it a traveling weather disturbance, reaching different places at different times in a sequence? What kind of thing was the forty- to fifty-day oscillation?

In the following year, 1972, Madden and Julian answered most of these questions in a second paper. In this one, they analyzed data from many radiosonde stations throughout the tropics, most of them still on islands. They found that the oscillation (which we now know as the Madden-Julian Oscillation, or MJO) was an enormous, planetary-scale disturbance. It occurred coherently, nearly in phase over very large regions. When one place was in the phase of the oscillation with westerly winds, so were other places near it. The opposite phase, with easterlies, would be occurring simultaneously at longitudes very far away. Then the phases would change, as the giant atmospheric disturbance moved steadily from west to east. It was strong but slow-moving in the equatorial Indian and western Pacific oceans. It sped up as it progressed into the central and eastern Pacific and Atlantic.

This was a huge discovery. But in the decade or so following these two papers by Madden and Julian, there was little additional research on the MJO. A few other scientists cited the 1971 and 1972 studies in their own articles, but many did so because of the statistical method rather than the scientific content. Tropical meteorology was still an intellectual backwater. The field was still dominated by the ideas that had grown out of Scandinavia earlier in the century, relevant only to higher latitudes. No one knew what to do with Madden and Julian's oscillation.

By the 1980s, however, the field had evolved. Tropical meteorology, though still a relatively raw area of research, had begun to mature, and more researchers became interested in the MJO. Other scientists began to use new data to reproduce and elaborate on Madden and Julian's findings; they confirmed all the features identified in the original papers. Theorists began to try to explain the MJO. They did not have much success. Most of us in the field don't believe that any of the early theories were correct. We are still arguing intensely about why the MJO exists and how it works. This is an active research area for me in particular.

We do now know the MJO to be globally important. There are only a few true, distinct phenomena in the climate system that can truly be called oscillations—meaning, periodic enough that they generate strong, clear peaks in power spectra of atmospheric variables. The MJO is one of those few. It's a signal that emerges above the meteorological noise.

The MJO is especially strong in the equatorial Indian and western Pacific Oceans, the adjacent continents, and the Indonesian region in between (known to climatologists as the Maritime Continent). These regions have monsoonal climates, meaning that their seasons are divided into dry and wet, rather than cold and warm, with a shift in the direction of the wind: the monsoon winds blow from the west in the wet season, from the east in the dry.

During the wet season, the MJO passes through, sometimes almost regularly every month or two, sometimes erratically, causing the weather to shift. The active phase is a wet period of a week or two, with strong west-to-east monsoon winds. In the suppressed phase, the weather becomes dry. It's as if the monsoon season were over, although it isn't. The monsoon winds weaken, or even shift back to blow from east to west. The skies are likely to stay clear until the next active phase starts to build, weeks later.

The wet and dry phases progress slowly from west to east, often starting over the Indian Ocean, then making their way east into the Pacific, where Madden and Julian first found them. Each active phase brings not an individual weather system, but instead a very large region of humid and disturbed atmosphere. Multiple rainstorms pass through over the span of a week or two. In between them, skies are overcast, with multiple cloud layers, stratus and cirrus, detritus from the last cumulus updrafts. The MJO is the "envelope" within which those storms exist. It is not exactly weather and not exactly climate, but in between.

The MJO has profound effects on weather over nearly the entire globe. It is the most important climate phenomenon you have never heard of.

The MJO disturbs the deep ocean. It causes the rotation of the earth to slow down and speed up as it goes through its cycles, with winds dragging against mountain ranges. The MJO makes the days a little longer, then shorter again. The changes are tiny (small fractions of a second), but large enough to be easily measured by modern instruments.

The MJO's impact is not only local to the tropics, where the MJO lives, but reaches up into the higher latitudes. The rainstorms in the active phase disturb the circulation of the atmosphere on a planetary scale. Enormous ripples called Rossby waves—named after the same Carl-Gustaf Rossby who ran the University of Chicago meteorology department—emanate into the jet streams of the midlatitudes. In Northern Hemisphere winter, these MJO-driven waves often cause floods on the West Coast of the United States. They tilt the jet stream so that it angles from the subtropical central Pacific northeast toward California, Oregon, and Washington. These "Pineapple Express" events (so called because the origin of the humid air is in the general vicinity of Hawaii) bring warm, humid subtropical air streaming into the high, cold mountains of the coastal ranges.

Back in the tropics, the disturbances that come with the MJO take on as many forms as tropical weather can. Among those are tropical cyclones. Tropical cyclones are much more likely to form in an active phase of the MJO than a suppressed phase. The atmosphere becomes very humid in the active phase, up to great heights and over a large area. This is just right for tropical cyclogenesis. Besides the right environment, tropical cyclones need "seeds," or preexisting weather disturbances, from which to grow. The smaller-scale storms that emerge in the active MJO phase provide many.

Although the MJO is most intense and vigorous in the Indian and western Pacific Oceans, it continues eastward beyond there. It passes through the eastern Pacific, across Central America, and into the tropical Atlantic. In the Atlantic, the MJO's influence becomes more subtle. It's clearly evident as a disturbance to the upper-atmospheric winds, but it doesn't cause as much rain and cloudiness. In 2000, though, my colleague Eric Maloney, then a graduate student at the University of Washington, in a paper with his PhD thesis adviser, Dennis Hartmann, showed that an active phase of the MJO strongly favors hurricane development in the Atlantic. It is *four times* more likely that a hurricane will form in the Gulf of Mexico or Caribbean during the active phase of the MJO than during the suppressed phase.

In just the last few years, this knowledge has begun to be used in fore-casting. Starting in 2013, the National Hurricane Center's Outlooks have in-cluded predictions of the likelihood of tropical cyclogenesis out to five days, instead of two. This is because of improvements in the computer models used for weather prediction, but also because of better understanding of the role of the MJO. By moving slowly, the MJO adds a predictable element to tropical cyclogenesis.

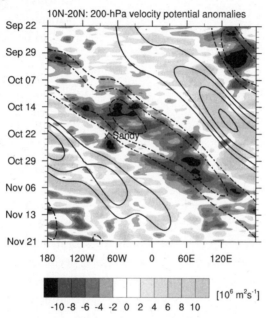

Plot showing the propagation of the Madden-Julian Oscillation (MJO) from the Pacific into the Atlantic during the period around Sandy's genesis. The shading shows the upper-level velocity potential, a quantity that is broadly indicative of the presence or absence of deep convection and disturbed weather. The contours show the same quantity filtered to isolate the signal associated with the MJO. The x axis is longitude, the y axis is time (running down), and the velocity potential data are averaged between 10 N and 20 N latitude. (This format, with time on one axis and distance on another, is known as a Hovmöller plot.) The dark shading and dot-dash contours angling down and to the right at the center of the plot indicate an MJO disturbance moving from the Pacific into the Atlantic. The time and longitude of Sandy's official genesis (when it was judged to have achieved tropical storm intensity and was named) are also indicated, showing that genesis occurred just after the passage of the MJO disturbance. (© 2012 Michael Ventrice)

In mid-October 2012, a newly developed active MJO phase was moving from the Pacific east into the Atlantic. The forecasters at the National Hurricane Center saw it coming, as did their colleagues at the Climate Prediction Center. They all knew that the environment in the Caribbean would become favorable for a hurricane. When an easterly wave ambled west into the active MJO phase there, they weren't surprised when, on October 22, this system intensified into Tropical Depression Eighteen, and then Tropical Storm Sandy just a few hours later. The easterly wave was the seed. The MJO, an immigrant from the Pacific, made the "ENVIRONMENTAL CONDITIONS . . . FAVORABLE" for that seed to grow.

The forecasters' job is to track these kinds of developments, but the situation didn't yet contain any obvious drama. While any hurricane forming in the Caribbean has the potential to be dangerous to someone somewhere, this late-season hurricane formation was not particularly unusual or remarkable. Some of the computer weather models being run at the same time, however, had a clear and true vision of Sandy's future.

3

MONDAY, OCTOBER 22

On Monday, October 22, at 1500 hours Universal Time (10:00 a.m. on the U.S. East Coast), NHC issued its first advisory on Tropical Depression Eighteen. This was the system that had been the tropical wave and was juiced up when the active MJO phase moved over it. Now it had moved up to "depression" status, a somewhat vague category meaning not quite a tropical storm, but getting close. Depressions get numbers, as they wait in line for the proper names they'll get if they intensify enough.

The text of the advisory was matter-of-fact. It stated that a tropical storm watch had been put in effect for Jamaica, and that either a hurricane watch or a tropical storm warning might be needed later that day. (A hurricane is stronger than a tropical storm, but a warning carries a higher degree of certainty than a watch.) The center of the storm was located near 13.5 N, 78.0 W.

In addition to its public advisories and forecasts, NHC issues "discussions." These are explanations of the forecasters' thinking, in technical language. The first discussion of Tropical Depression Eighteen, issued at 11:00 a.m. Eastern Daylight Time on October 22, by forecasters Lixion Avila and Robbie Berg, began its final paragraph (ellipses in original):

THE DEPRESSION IS LOCATED IN AN ENVIRONMENT THAT IS HIGHLY CONDUCIVE FOR STRENGTHENING . . . AT LEAST DURING THE NEXT 36 TO 48 HOURS. IN FACT . . . THERE IS A 50/50 CHANCE

THAT THE DEPRESSION WILL STRENGTHEN BY AT LEAST 25 KT DURING THE NEXT 24 HOURS BASED ON THE RAPID INTENSIFI-CATION INDEX. THEREFORE ... THE OFFICIAL FORECAST SHOWS FAIRLY QUICK STRENGTHENING DURING THE NEXT 48 HOURS ... AND THE DEPRESSION COULD BE ON THE CUSP OF BECOMING A HURRICANE AS IT IS APPROACHING JAMAICA.

This thinking would remain consistent in the forecasts over the following days as the storm developed. It would be verified as correct.

These sentences also indicate much about how good hurricane forecasters think. The storm was believed likely to intensify not because of anything it was doing itself, but because its environment was "conducive for strengthening." The environment means the conditions of the atmosphere in a larger area around the storm, represented by measurements from satellites and weather balloons launched from nearby islands. This is nurture rather than nature; nature plays a role, too, but nurture is much easier to observe, understand, and predict.

The environment was being disturbed by the MJO. The giant disturbance from the Pacific was lifting the atmosphere, moistening it, and preconditioning it for cumulus clouds. It was also shifting the winds in such a way as to keep vertical shear (the difference between the winds at different heights) small, reducing the chance that the incipient storm's top would be torn away from its bottom.

The "rapid intensification index" that the discussion mentions tells us something as well. Rapid intensification is just what it sounds like: a storm becoming much stronger very quickly. It can be a nightmare for forecasters, because the computer models they rely on aren't any good at predicting it. So they use an index. That is a number that measures a combined set of factors (properties both of the storm itself and of the larger environment around it) that have been associated in past storms with rapid intensification. Compared to the computer models, the index is a crude tool. But despite the current models' great sophistication, and their considerable skill in predicting hurricane tracks, they are still poor at predicting intensity. The index doesn't work much better, but the fact that it is used at all is an indicator that intensity forecasting is way behind the other aspects of hurricane prediction.

The discussion continued:

AFTER 48 HOURS . . . VERTICAL SHEAR IS EXPECTED TO INCREASE SUBSTANTIALLY . . . AND NEARLY ALL THE GLOBAL MODELS SUGGEST THAT THE CYCLONE WILL TAKE ON MORE HYBRID CHARACTERISTICS . . . SUCH AS AN EXPANDING WIND FIELD TO THE NORTH AND ITS INTERACTION WITH A MID- TO UPPER-LEVEL LOW.

"Hybrid" means having some properties of an extratropical (or winter) storm. The word's use here implied the "expanding" wind field: strong winds over a larger area than typical for a hurricane. All this, too, would prove correct.

Just six hours later, the next advisory on the system upgraded it to Tropical Storm Sandy. The name, as every year since 1979, was taken from an alphabetical list prepared before the start of the Atlantic hurricane season, in which masculine and feminine names alternated.[7] Sandy, a name given to human beings of both genders, was a feminine name this year, sandwiched between Rafael and Tony. The storm's maximum sustained winds were estimated at thirty-five knots. This is the typical value for a newborn tropical storm; below that, it wouldn't be a tropical storm at all, but still a depression. Central pressure was 999 hectoPascals (given in the advisory in the equivalent units of millibars, and abbreviated mb). A moderately depressed value, consistent with the thirty-five-knot winds, and reminiscent of a price on an item of clothing in a bargain outlet store.

With the intensification, the geographic area threatened by Sandy began to expand. Tropical storm watches were declared for both Jamaica and Haiti, and forecaster Richard Pasch wrote that "INTERESTS IN EASTERN CUBA AND THE BAHAMAS SHOULD MONITOR THE PROGRESS OF SANDY." At the moment, though, the storm wasn't going anywhere. Oxymoronically, the advisory stated, "Present movement is stationary."

The advisory forecast that Sandy would drift northward over the next several days, while intensifying gradually. The maximum sustained winds were forecast to reach sixty knots (just under the bar for being labeled a hurricane) by 1800 Universal Time (UTC[8]) on Wednesday the twenty-fourth, decreasing again to fifty knots on Thursday, at which time it was predicted to have reached 22 degrees north, 76 west—just shy of the Tropic of Cancer, between Cuba and the Bahamas. The extended outlook for days four and five, Friday and Saturday, had Sandy staying at fifty knots, then decreasing to forty-five, while continuing to make its way north and yet farther east, keeping it far from the U.S. coast.

NHC forecasters make their assessments of a storm's current state based on a suite of observations: from several satellites, sensing electromagnetic radiation in the visible, infrared, and microwave bands; from surface observations and upper-air weather balloon measurements from nearby islands; and from anything else available. That may include surface observations from nearby ships or from weather-observing buoys on the open ocean. When a storm starts to appear threatening, the U.S. Air Force and NOAA fly hurricane reconnaissance aircraft into and around the storm for a closer look. Sandy hadn't yet reached that point.

All the same data the forecasters are looking at, plus many more data from around the globe, are continually fed into computer models. The models are run at numerical weather prediction centers around the world. Each model ingests all the data and uses them to estimate the state of the global atmosphere. That state is the starting point for a projection forward in time, produced by solving digitized forms of the mathematical laws of physics. The resulting solutions are the basis for all modern weather forecasts, including hurricane forecasts. The NHC forecasters base their predictions, nowadays, largely on the guidance they get from these models.

The American model is run on computers at NOAA's National Centers for Environmental Prediction, in Maryland. In October 2012, NCEP had just moved into a new building in College Park. NCEP's model is called the Global Forecast System, or GFS. The NHC five-day track forecast for Sandy, issued on October 22, taking the storm out to sea, was close to the GFS model's solution at that time.

NHC's forecasts do not go beyond five days, but the models are run longer than that. At the longer lead times, the errors in the model solutions normally grow larger, and forecasts based on those solutions become less reliable. They are much better than they used to be, though. A seven- or ten-day weather forecast now has some skill in it. You shouldn't make a big bet on any one forecast that far ahead, but if you were to bet on many of them, over time, you would make money. Many other forecasters, in and out of government agencies, do look at the long-range model solutions. They are visions of the future, as consistent as we can make them with both the present moment's weather and the rules by which nature causes the weather to change from one moment to the next.

The GFS's vision was that, beyond five days, Sandy would continue both north and east, away from land. Other models, including that of the European Centre for Medium-Range Weather Forecasts (ECMWF), run in Reading, UK, and the Canadian model, run by Environment Canada, had a different vision. They saw, in essence, *exactly what would eventually happen eight days later.*

For the first few days, the European and Canadian models took the storm north and a little east, very close to the GFS track. All the models, at that point, showed it beginning to expand into a larger storm as it interacted with disturbances in the higher-latitude autumn jet stream. At a critical juncture, though, the European and Canadian models zigged where the GFS model zagged. They predicted that the storm, after paralleling the coast for some distance, and while growing still more in size and intensifying further, would take a left turn instead of a right, and hit the coast head-on. They forecast that it would make landfall somewhere in the mid-Atlantic as an enormous, powerful storm. According to the Canadian model run on October 22, the landfall point would be in southern New Jersey.

Most kinds of quantitative predictions that human beings make (predictions of the stock market, elections, or baseball games) are based on statistics.[9] If you want to know whether something will happen, look at what other things were going on the last time it happened, use some judgment to decide which of those things were important, and see if those same things are going on similarly now. Past performance may not predict future results, but in most situations it's all we have.

Weather prediction is nearly unique in that we do not do it this way. We do it based on knowledge of the underlying dynamics of the system. We know the real rules that the atmosphere is bound to obey. These are the laws of physics. Sometimes those laws lead to outcomes that human beings have never witnessed. Our models, in principle at least, can still envision them. The outcomes are consequences of those laws, encoded into the models.

Sandy was such a case. No known Atlantic hurricane had ever made a left turn into the northeastern United States like the one the models envisioned on October 22. If any person had tried to predict the track statistically, it would have been nearly impossible for that person to make a correct forecast.

The October 22 formation of Tropical Storm Sandy, soon to become Hurricane Sandy, was not particularly unusual. Nor would the first few days of Sandy's life be anything special, by historical standards, though the storm would be destructive as it passed through the Caribbean. But the models saw that as the MJO had been moving into the Caribbean to spark the tropical wave's intensification, the polar atmosphere had also been shifting its gears. Two enormous eddies were locking into position in the upper air over the northeastern United States and the western North Atlantic.

4

NAO

I am writing this chapter in Paris, in late June. I have the pleasure of being a visitor to the Laboratoire Météorologie Dynamique, in its location at the Ecole Normale Supérieure, near the Luxembourg Gardens. I've been here a little over a week, and the last time the temperature rose above 20 degrees Celsius (68 degrees Fahrenheit) was five days ago. The first half of the month, before I arrived, was dry. Then, enough rain fell in my first few days here to put the monthly total above average. In short, it hasn't been particularly summery. I asked my colleague and host, Jean-Philippe Duvel, if this time of year was often this cool and gray. He said, "No, it's the NAO."

The climate of Europe and the North Atlantic is famously fickle. The weather can change suddenly and dramatically. It can also stay locked in an anomalous, unseasonal configuration for a long time. Since at least the Middle Ages, some observers have recognized that these variations don't grip the entire region uniformly. In fact, there are recurrent patterns in which certain specific places tend to experience one kind of weather while, simultaneously, specific other places experience the opposite.

As early as the thirteenth century, Norse settlers in Greenland observed that unusually cold but clear weather there tended to occur simultaneously with storms in nearby regions.[10] By the nineteenth century, with more modern observations and better scientific methods, the patterns of North Atlantic climate variability began to come into clearer focus.

The jet stream has a tendency to shift between a more southern and more northern position. The north–south shifts are indicated on maps showing the typical pressure patterns associated with the opposing phases of the North Atlantic Oscillation, the NAO that Jean-Philippe mentioned. The jet stream defines the storm track, the path along which storms move. When the jet shifts north, the storms follow it, bringing low pressure to northern Europe. The southern latitudes of the Mediterranean region tend to have high pressure and clear skies. That's the positive phase of the NAO. When the jet shifts south, in the negative phase, the storms follow. Southern Europe and the Mediterranean are under low pressure, and wet, while the north becomes clear and cold, with high pressure.

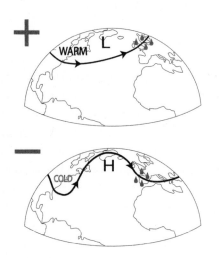

Schematic indicating the typical configuration of the jet stream during the positive (top) and negative (bottom) phases of the North Atlantic Oscillation, or NAO. The "H" in the negative phase indicates the frequent occurrence of blocking highs while the "L" in the positive phase indicates the strengthened low pressure over Greenland. Raindrops show the location of greater European storminess in each phase, while warm and cold anomalies are indicated over the eastern United States. *(Courtesy of Marit Larson)*

The NAO shifts are strongest in the winter, but can occur year round. Their patterns are recognizable, but they are never exactly the same. Right now, as I write, the NAO is in its positive phase, with a northward-shifted jet. That would normally bring warm, dry weather to Paris, but instead it's been stormy, and now cold. We're between high pressure moving in from the west and a slow-moving low-pressure disturbance over the Baltic Sea that has been bringing cold winds down from the northwest for days now. In the jargon of statistical climatology, we say that the NAO does not "explain a large fraction of the variance," meaning that knowing the state of the NAO doesn't tell us with great precision what the weather will be in any particular place. This is especially true in summer, when the NAO's influence is usually weak. My colleague Jean-Philippe's comment about the NAO was both right and wrong—there is a significant NAO event in place, but it doesn't explain my cold and gray Parisian welcome.

Actually, the NAO is an Atlantic manifestation of an even larger pattern that plays out over the entire Northern Hemisphere. The jet and storm tracks can shift north or south at any or all longitudes. The global version of the phenomenon is now called the Arctic Oscillation (AO). David Thompson and Mike Wallace of the University of Washington—the same Mike Wallace who was analyzing tropical radiosonde data in the late 1960s, just before Madden and Julian discovered their oscillation—named it that in their journal article from 1998.[11]

While the MJO deserves the name "oscillation," the NAO really does not. Many scientists have performed spectral analyses on the history of the NAO's flips and flops, looking for preferred frequencies. No one has convincingly found one. The NAO can change states rapidly, from one week or even one day to the next, but it can also stay locked into one state for longer times. There have been entire decades, or periods of several decades, where it seems to prefer one state over another. Then it shifts the other way, slowly or abruptly. The NAO emerges spontaneously from the chaotic "noise" of the weather in middle and high latitudes, with its jet streams, low- and high-pressure systems, and sharp temperature fronts. The global atmospheric flow, especially outside the tropics, is turbulent. The particular patterns of the NAO form out of that turbulence, stay for a while, and reverse, unpredictable even though sometimes long-lived.

Whether the NAO or the AO, it has an influence that reaches from Europe across the Atlantic and into North America. Just as over Europe and the Atlantic Ocean, the jet stream can shift north or south over the United States. Extratropical, or winter-type, storms (the generic type that occur in midlatitudes, sometimes so called even when they occur in summer) ride along the jet. In the positive phase of the NAO, the jet is north. In the negative phase, it's south, and storms dip farther south with it, over the central and eastern United States. Cold air outbreaks from Canada plunge down behind cold fronts that trail each storm.

In mid-October, as the MJO was moving into the Caribbean, the NAO was heading into its negative phase. By October 22, when Tropical Storm Sandy formed south of Cuba, the value of the NAO "index" was close to minus two standard deviations. Unlike the rapid intensification index used for hurricane prediction, the NAO index is just an indicator of what is happening now. Just like a stock market index that gives a rough measure of the performance of a whole category of securities, a climate index of this type measures the strength of a particular pattern. "Minus two standard deviations" is statistical language for "quite strongly in negative territory."

With such a strong negative NAO, the jet stream had begun to dip quite far south over the eastern half of the United States. This pattern had not been present in early October, and hadn't been predicted long in advance, but the models were now showing it would hang around for a week or more. On October 22, the CPC's six- to ten-day Outlook for the United States (a different product from the Global Tropics Hazards and Benefits Outlook, but produced by the same institution) was written by forecaster Randy Schechter, and began:

6–10 DAY OUTLOOK FOR OCT 28—NOV 01, 2012

TODAY'S ENSEMBLE MEAN MODEL SOLUTIONS ARE IN GOOD AGREEMENT ON THE EXPECTED 500-HPA FLOW PATTERN FOR THE 6–10 DAY PERIOD OVER NORTH AMERICA. A RELATIVELY HIGH AMPLITUDE REGIME IS ANTICIPATED OVER THE CONUS, WITH TROUGHS FORECAST OVER THE EASTERN CONUS AND EASTERN ALASKA WHILE RIDGES ARE ANTICIPATED OVER THE NORTHWEST ATLANTIC AND WESTERN CONUS.

"High amplitude regime" meant the jet stream was very disturbed, with a large undulation in it. CONUS refers to the continental United States. A "trough" means low pressure; a "ridge" means high. The Outlook went on to state that the different weather models were computing quite similar predictions for the upcoming period, starting six days from then, on October 28, and lasting through November 1.

The CPC six-to-ten-day forecast map, depicting the height of the 500 hPa pressure surface, is shown in the color insert. The height at a given pressure means about the same thing as the pressure at a given height. This map basically shows the pressure field about three miles up.

If the planet were not rotating, significant differences in pressure from one place to another would not last long. Air would flow from high pressure to low, and the pressure field would flatten out; maps such as the one shown in the CPC forecast in the color insert would show almost nothing. But because the earth is rotating rapidly, the air currents are always being turned sideways, as they try to move from high to low pressure, by the Coriolis force. The Coriolis force is an apparent force that results from something's being in a rotating frame of reference. Think of trying to play catch with someone standing opposite you on a fast merry-go-round. As the ball moves from you to your partner, she rotates out of the way. But from where she stands, the ball appears to fly in a curved path away from her. The Coriolis force is the apparent force that causes that curve.

In high latitudes, the earth's rotation axis points nearly vertically, through the planet's surface, like the axle of the merry-go-round. The Coriolis force is strong. Instead of flowing from high to low pressure, it turns the wind so that it flows sideways between the two, along the isobars, or lines of constant pressure. This is called "geostrophic" wind. When one looks at a map such as the one in the CPC forecast in the color insert, the contour lines define the streamlines along which the air is moving, approximately, because the wind is close to geostrophic. If the wind were a hiker and the contours showed elevation on a trail map, the hiker would be traversing the slope, avoiding going uphill or down. Because of the direction of the earth's rotation, and because the low pressure is closer to the pole while the high pressure is closer to the equator, the air in this particular map is moving roughly from west to east.

On its way from west to east, though, in the prediction made on October 22 for the period October 28–November 1, 2012, the air was taking a large detour. The contours dip down to the south over the eastern United States, indicating air moving southward over the Midwest on the west side of the upper trough, and back to the north along the coast. This trough itself was not exactly what we would call a storm. It was a larger-scale "planetary wave" (also known as a Rossby wave) in a forecast of the five-day average flow. In a flow pattern like this, smaller storms—their typical size maybe several hundred miles wide, what we call synoptic scale—typically develop along this large trough and move through it, strengthening on the east side.

Farther east, just offshore of the Canadian Maritime Provinces and northern New England, the contours bulge far north, indicating air moving clockwise around an area of high pressure south of Greenland.

When an area of high pressure stays in one place for more than a day or two, and that place is located in a storm track, meteorologists call that a blocking high, or just a block. The storms move from west to east, following the average winds in middle and high latitudes. When the storms come up against a block, they can't go through it; they either stall in place, or go around it. Since the map in the CPC Outlook represents an average over a five-day period, the northwest Atlantic high-pressure system shown there is a block by definition. It was setting up already on October 22, as Tropical Storm Sandy was forming in the Caribbean. The CPC forecast was that the block would remain in place long enough for Sandy to run into it on the poleward track that the models were beginning to envision.

In late October 2012, as the NAO index was diving far into negative territory, the pattern over the eastern United States and western Atlantic was both typical and atypical of the negative NAO pattern. A deep trough over the eastern United States is a typical feature of a negative NAO event. Cold air at the surface comes down from Canada behind the trough, and this, too, was in place.

While a blocking high over Greenland is a typical feature of the NAO, the block in late October 2012 was not typical. It was not over Greenland, but well south of it. Rather than being just an obstacle in the way of storms at high latitudes, this block was close enough to the northeastern United States to be within striking distance of any storm that might come north, paralleling the coast, from the tropics.

5

TUESDAY, OCTOBER 23

Sandy started its life as a named storm slowly. In its first twenty-four hours after becoming a tropical storm, it didn't move far, or strengthen much. Between 5:00 p.m. Eastern Daylight Time on the twenty-second, when Sandy was first declared a tropical storm and given its name, and 8:00 p.m. on the twenty-third, NHC's estimate of the maximum sustained winds increased from thirty-five knots (the threshold to name a storm) to forty-five. Not a dramatic increase, but enough to give credence to NHC's forecast that Sandy would likely become a hurricane soon. During this same period, the storm moved around one hundred fifty miles to the north and seventy or so to the east.

The 11:00 a.m. forecast discussion on the twenty-third, by forecaster Richard Pasch, began:

THE CLOUD PATTERN OF THE STORM IS GRADUALLY BECOMING BETTER ORGANIZED . . . WITH A LARGE CURVED BAND FEATURE EVIDENT OVER THE EASTERN AND SOUTHERN PORTIONS OF THE CIRCULATION. USING A BLEND OF SFMR AND FLIGHT-LEVEL WIND DATA FROM THE HURRICANE HUNTERS . . . THE CURRENT INTENSITY IS SET TO 45 KT.

The "Hurricane Hunters" are the U.S. Air Force scientists who fly through hurricanes, in C-130 military transport airplanes that have been retooled with meteorological gear. The Hurricane Hunters had begun flying the storm on the previous day, and would continue throughout Sandy's lifetime. They would be joined later by the other U.S. aircraft used regularly for hurricane reconnaissance, the P-3 research planes flown by NOAA.

Visible satellite image showing Tropical Storm Sandy at 20:45 UTC (4:45 p.m. Eastern Daylight Time) on October 23, 2012. At this time the center was near 14.5 N, 77.5 W, in the Caribbean sea. Image courtesy of NOAA, from http://rammb.cira.colostate.edu.

The data these flights provide can be invaluable for forecasts. Their utility begins with more precise estimates of a storm's present intensity than any other observations can provide. There are typically no other measurements from directly inside the storm. Satellites are the backbone of hurricane forecasting globally. They are always there, and collectively they can see the whole planet. Pasch's observation of better organization and curved bands in the clouds came from satellite images. But no satellite can measure the winds directly from space.

Even with the advantage of an airplane directly inside a hurricane, a precise estimate of a hurricane's present intensity (a prerequisite for a good forecast of its intensity in the future) is not a simple matter. The SFMR mentioned in the discussion refers to a Stepped Frequency Microwave Radiometer, an instrument that looks down at the ocean's surface. It measures radiation emitted by the sea over a range of frequencies in the microwave range—"stepped" means it does one frequency at a time. The SFMR estimates the strength of the surface winds from above by sensing the roughness of the ocean's surface. The plane also has instruments that measure the wind directly at the plane's location; the "flight level" is typically a couple of miles above the surface. This is a direct measurement, but not exactly where it's most wanted; the wind at the surface of the ocean is what's used to categorize the storm's intensity. Hence the "blend" of SFMR winds (at the ocean's surface, but not a direct measurement) with the flight-level winds (direct but not at the ocean's surface).

Sandy's predicted track for the near term took it to the north-northeast, with its pace forecast to pick up. NHC's line on the map showed the next five days' path with its standard cone of uncertainty, indicating the typical range of errors in the track forecast at each lead time. The farther ahead in time, the more likely, based on the errors in all forecasts made over the last five years, the forecast position will miss the mark by a greater margin. The cone of uncertainty would prove conservative in Sandy's case; the track predictions, unusually accurate.

For now, in the next day or two, the forecast had Sandy making landfall in Jamaica, then likely in eastern Cuba, as a hurricane. The sequence of NHC advisories from this time shows a chain of watches and warnings beginning to cascade northward across the Caribbean, from Jamaica, into Haiti and Cuba, then up to the Bahamas: a tropical storm watch becomes a tropical storm warning, often coinciding with a hurricane watch, which then turns into a hurricane warning, as the possible future danger turns into a likely and imminent one.

Early in the morning on October 24, Sandy became a hurricane. Observations from another Hurricane Hunter flight indicated maximum sustained surface winds of sixty-five knots. Satellite images from 1200 UTC (0600 EDT) show an eye at the center. The storm was about one hundred miles south of Jamaica, with the pace of its motion to the north quickening. Its first landfall occurred in Jamaica seven hours later, in the early afternoon. The NHC report puts the center at landfall near Bull Bay, between Kingston and South Haven, and the intensity as seventy-five knots, a category one hurricane.

Though Jamaica is right in Hurricane Alley, it had been lucky for some time. Sandy was the first hurricane to hit the country head-on since Gilbert, twenty-four years earlier.[12] Compared with what could easily have happened (had Sandy been a stronger storm at landfall in Jamaica, as it would be shortly after), the country was, still, fortunate. The winds tore corrugated metal roofs off homes, and closed the airports for around a day, and roads for longer.[13] Most of the island lost electric power. Crops were lost, as winds knocked down banana trees and coconut palms.[14] But only one person died in the storm.

Haiti, to the east of Jamaica, was not directly in Sandy's path, as defined by the line on the map showing the motion of the center. But it was close enough. The southern part of the island was drenched by rain; some gauges measured twenty inches. Few places on earth would have been more vulnerable to this. Already one of the world's poorest countries, Haiti had been devastated just two years earlier, in 2010, by a massive earthquake. Many people were (and are) still living in tents. With the landscape deforested (most trees cut and sold for charcoal), the soil had washed away easily, and there was little to slow down the water rushing into overflowing rivers. A few days after the storm, the United Kingdom's *Guardian* reported a death toll in Haiti of fifty-two; if this disaster was similar to others, it is likely that more people died later of causes indirectly related to the storm, such as disease.

Hurricanes get their energy from the ocean, and weaken rapidly upon landfall. But Sandy completed its transit across Jamaica too quickly to lose much strength over the island. Reemerging after just a few hours into the Cayman Trench, the narrow channel separating Jamaica from Cuba, the hurricane resumed its aggressive strengthening. Sandy's center took only a few hours to cross the water, making landfall in eastern Cuba in the middle of the night (0525 UTC on the twenty-fifth, per NHC). In that short time, its winds dramatically intensified. Though called a category two at the time, after-the-fact analysis would raise that to category three, with one-hundred-knot sustained winds.

The Saffir-Simpson hurricane scale, the grading scheme from which the categories of hurricane intensity derive, is deceptive in several ways. Most of all, it measures only wind. We who lived through Sandy in the north-eastern United States know from hard experience that wind is not always the primary threat. In many places, storm surge is the weapon by which hurricanes destroy life and property. In others (Haiti during Sandy being one), flooding rains cause the catastrophe. But even in cases where wind is the threat, the ranking by categories one through five doesn't do justice to the rate at which damage skyrockets as a storm makes its way up the scale.

Damage due to Sandy in Santiago de Cuba. *(Courtesy of Desmond Boylan/Reuters)*

The Saffir-Simpson scale is just a set of ticks on a ruler that measures the maximum sustained winds. In fact, this is another problem; in ranking a storm only by the peak sustained wind speed, the scale doesn't tell us anything about the spatial extent of the winds. How large an area will experience either those maximum winds or winds that are a little less than the maximum, but still strong enough to be destructive? Even leaving that aside, the scale does not hint at *how rapidly* the damage done by wind increases with wind speed. Just how rapidly is not entirely settled by the scientists and engineers who study the question, but damage is proportional at least to the wind speed *cubed* (and maybe a higher power still). A hundred-knot storm (a low category three) has winds that don't sound that much faster than those of a seventy-five-knot category one storm. If the speed limit on a highway were seventy-five, you might be able to do one hundred for a while without getting a ticket or causing an accident. But the hundred-knot storm's winds are at least two and a half times as dangerous as the those of the seventy-five-knot storm, maybe more.

Cuba was battered by Sandy. The center made landfall close to the city of Santiago de Cuba, the country's second-largest city, and wrecked it. Estimates of the number of homes damaged are in the six figures, some as high as two hundred thousand; as many as fifteen thousand were entirely destroyed. Power outages were widespread, some lasting weeks.

Eleven people were killed by Sandy in Cuba. This is a larger number than in Jamaica, but a much smaller one than in Haiti, where Sandy was weaker. This is not an accident; it's our first indication that a natural disaster is not really natural. Nature throws the switch, but what makes us register it as a disaster is what happens after that, to the people and their stuff. This depends on who those people are and how they are set up physically, economically, and politically.

6

DEADLY BORING: THE HURRICANE PRONE

Let's be frank. Sandy became a big deal because it hit New York City. If, after passing through the Caribbean and making its way north offshore of the continental United States, it had turned right instead of left, and headed out to sea to die quietly over the North Atlantic, I wouldn't be writing this, and even if I had written it, you probably wouldn't be reading it. The damage Sandy did in Jamaica, Haiti, Cuba, and the Bahamas would have been a minor news item, for a couple of days at most, in the United States and the rest of the world, and that would have been it.

I don't say this as a criticism of either you or me. It's the nature of being human and living in the world we do. We pay attention to things when they affect us directly; when they have a large impact on some part of the world that means something to us; or, failing both of those, when they are so unusual, so extreme, that they can shock us into attention nonetheless. What happened to New York, New Jersey, and the surrounding areas when Sandy hit was unprecedented for the city and the region. Such a disaster, in its combination of the natural event and its impacts on this place, existing now in its modern, metropolitan configuration, had truly never happened before. Because of the national and global importance, and visibility, of the New York City metropolitan area, Sandy became a story with meaning for the world.

We might also point out that if measured in dollars, the damage done to the Caribbean was minuscule compared to that inflicted a few days later on the United States. Of course, if measured relative to the sizes of the respective nations' economies, the financial losses in the Caribbean balloon to much more massive proportions relative to those in the United States. But then, to be less cynical again, we might point out that those Caribbean countries have small populations compared with that of the United States. If we measure importance by the number of people affected in some way by the storm, the United States again looms largest by far.

Surely, though, a large part of the reason that our attention isn't seized by a hurricane-induced disaster in the Caribbean, whether Sandy or any other, is simply that it happens all the time, and always has. The 2010 Haiti earthquake, as a counterexample, was a fairly major U.S. news story—not as major as Sandy, but pretty major, considering how little appetite our national media have altogether for anything outside our borders. Yes, that earthquake was a huge disaster by any measure, but it was also unusual. A hurricane, down there, just isn't.

The problem is well known to scholars of the region. Historian Stuart B. Schwartz, in an article on the shaping of the Caribbean by hurricanes since European settlement,[15] summarizes it vividly:

> The writing of the history of hurricanes like that of much environmental history begins with a problem. For all their power and destructive potential, the history of the hurricanes is, because of their frequency, almost inherently boring. Unlike volcanoes or earthquakes, the great storms are somewhat dependable. Almost every year some island or coast is inundated or devastated. While for individual islands or cities or stretches of shoreline, hurricanes may be spaced decades apart, in a regional sense, the phenomenon is repetitious and the results expected.
>
> The scenes of destruction are all too common and all too similar: shattered homes and shattered lives, boats piled up on the beaches or carried far inland, destruction in all directions, posterior scenes of relief and aid amidst a backdrop of ruin. The individual stories may be poignant, but their repetition is numbing.

Caribbean hurricanes may be old, familiar news; they may give us disaster fatigue; but that doesn't make them any less disastrous. Their regularity has made them a constant influence on the peoples of the Caribbean throughout their histories. The long-term influence of persistent natural disasters on the development of societies is a profound, but still relatively little-studied subject. It deserves its own book (or several). I'm not the right person to write that book, but before we follow Sandy up the coast on its path toward New York, we can spend a short time on it.

The word *hurricane* comes from *Hurakan*, in the language of the Taino, one of the native Arawak peoples of the Caribbean. The word seems to have referred just to the storm itself. The Taino goddess Guabancex controlled the storms, with the help of two assistants, Guataba and Coatrisquie. Some images of Guabancex show her with outstretched arms forming a spiral. The spiral symbol used on weather maps today for a tropical cyclone appears to be inherited from the Taino. The Taino were wiped out as a distinct people (though some of their descendants, intermarried with the Spanish and others, are still in Puerto Rico, the Dominican Republic, and elsewhere) shortly after European settlement. If they knew that the hurricane's winds formed a spiral pattern, they knew it more than five hundred years ago. Western meteorologists did not know this until the nineteenth century.

When Europeans arrived in the New World, hurricanes were new to them. On Columbus's fourth voyage, in 1502, he met a hurricane off the coast of Hispaniola (the island now divided between Haiti and the Dominican Republic). Francisco Bobadilla, rival to Columbus for the favor of the Spanish Crown, had recently finished his term as royal governor of Hispaniola, and set sail with a fleet of twenty ships to bring treasure back to Spain. Columbus had seen the storm coming, and warned them, but they set out anyway; all ships, and all the men aboard them, were lost, while Columbus's own ship weathered the storm safely in a natural harbor. According to historical accounts, Columbus could forecast the storm by the "high cirrus clouds and southeasterly swell"; if so, this was a remarkable forecast, particularly as it is not clear that Columbus himself had actually experienced a hurricane in any of his previous voyages. In his first voyage, in 1492, he apparently did not encounter one; this was lucky, because he was in the tropics that year during September and October, peak hurricane season.[16]

The Spanish, and the other colonial powers that competed with them for influence in the Caribbean, subsequently came to know hurricanes well. The storms would sink many more treasure ships, and decide some military campaigns. The city of Havana was built where it is, on the north coast of Cuba, because hurricanes tend to strike from the south. The Spanish developed a regular schedule for sending Mexican and Peruvian silver and gold back to Spain: it was timed carefully so that fleets would exit the Caribbean to the north by June or July, avoiding the peak hurricane season, and then catch the midlatitude westerlies for the rest of the voyage across the Atlantic. The regularity of the schedule made the ships vulnerable to pirates and enemy fleets, but the storms were a worse danger.[17]

The colonies on a number of occasions had to petition their respective crowns (not only the Spanish, but the British and others) for financial help after hurricanes struck. This was the beginning of long-distance natural disaster relief. Often relief came, but sometimes it was not enough. The unsuccessful revolution of Puerto Rico in 1868, and the successful one in Cuba a few decades later, both were influenced by inadequate responses to hurricanes from the colonial powers.

The storms keep coming. Each one brings another disaster story. We can see their influence everywhere, if we look—not only in the Caribbean island nations, but in the parts of the United States regularly hit by hurricanes. Houston and Galveston were once rivals for dominance of the Texas economy. Houston blew up because of oil, true, but Galveston's trajectory was also abruptly halted when it was destroyed by a hurricane that killed many thousands in 1900.[18]

These are just Atlantic hurricanes, which make up only a little over 10 percent of tropical cyclones globally. The tropical-cyclone-prone regions of the western Pacific (comprising the entire southern Asian coastline and the islands of Japan, the Philippines, Taiwan, and others offshore) have their own deep histories with storms. Australia and Madagascar have had their catastrophic experiences with Southern Hemispheric tropical cyclones, which spin clockwise rather than counterclockwise, as ours do. Darwin, Australia, a small city, but by far the largest one for thousands of miles in any direction from its location at the continent's northern tip, was obliterated in 1974, with little warning, by Cyclone Tracy. Tracy was an intensely powerful storm, but also quite small; a slight change in its track would have been enough to spare Darwin. No such luck.

The South Asian nations bordering the North Indian Ocean experience fewer storms, but those they do have can be unbelievably awful. Bangladesh lies in an enormous flat river delta that opens to the Bay of Bengal. Large populations of people live in rickety houses in low-lying coastal areas vulnerable to the overwhelming storm surges that a powerful cyclone can produce. A single storm in 1970 killed at least 300,000 people in Bangladesh; another in 1991 killed 138,000. I did not type those numbers incorrectly; nor did you misread them. Those are numbers in the hundreds of thousands, each from a single storm.[19]

In recent years Bangladesh has developed much better early warning and preparation systems, with elevated shelters to hold evacuees. The most recent cyclone with the potential to be a major disaster was Sidr, in 2007; thanks to the improvements, the death toll from Sidr was around 3,500. For Bangladesh, when listed next to the 1970 and 1991 storms, that figure looks like an incredibly good result. But we shouldn't lose perspective entirely. The official death toll from Hurricane Katrina, considered an epic national catastrophe here in the United States, and a failure of government at every level from local to federal, was around half that. Life is still cheaper, apparently, in some places than in others.

How do natural disasters affect societies over the long term? The short-term impacts are obvious enough, but are the long-term impacts just the sum of those? Or do the responses of people and governments actually reverse those impacts, and turn disaster into opportunity? Does that which doesn't kill us make us stronger?

In the age of data science, it is possible to answer this question quantitatively. My colleague (and former student at Columbia University) Solomon Hsiang, with collaborator Daijung Narita, has studied it using a unique combination of economic data, records of the number of tropical-cyclone-related deaths, and a careful reconstruction of the physical properties of tropical cyclones at landfall. By studying 233 countries around the world, the two researchers were able to draw general conclusions about how societies and their economies respond to hurricanes, both in the short and long term. They studied differences among countries with different degrees of exposure to cyclones, and also variations in time, looking at economic output before and after landfall, as a function of the size of each country's economy (measured by gross domestic product), as well as fatalities, to see the storms' impact.

Their basic finding was that tropical cyclones impose a long-term cost.

That which doesn't kill us, if it's a hurricane, makes us weaker, just as the most naïve expectation might have it.

There's a little more to it than that. The short-term economic loss from the same storm is less in a country where storms are a routine occurrence than in one where they are a freak of nature. The interpretation of that finding is that hurricane-prone societies learn to adapt. They invest in things that help protect them from storms. The analysis doesn't tell us which things; we might guess that physical systems such as levees and more storm-proof buildings might be involved, or human ones such as better forecasts or early warning systems.

But the adaptations reduce costs only by a little. The difference between the economic hit from a storm in a highly adapted country and that in a poorly adapted one is only about 3 percent. The data offer no support for the view that the response to disasters strengthens economies overall. There may be a short-term boom in some economic sectors locally (construction, for example, as homes and businesses need to be rebuilt), but in the aggregate, hurricanes impose a cost on their victims that doesn't go away.

Hsiang and Narita's analysis also shows that the damage from the same hurricane is less in a higher-income economy than in a lower-income one. That's with the economic damage normalized by GDP. In absolute dollar terms, it's much easier to lose a lot of money in a rich country than a poor one, but proportionately the poorer country will often lose more.

In terms of deaths, the difference between Cuba's experience of Sandy and Haiti's is stark. Cuba was hit by a category three hurricane and lost eleven people; the major population centers in Haiti didn't experience even tropical storm–force winds, and lost fifty-two. Rain did deluge Haiti, but it wouldn't have been as deadly if so many people there hadn't been living in precarious shelters, on a deforested island, with weak institutions.

Cuba has an extremely organized and structured system of emergency preparation, warning, and response for hurricanes. Many branches of federal and local government are involved, each with well-defined roles. "Civil defense," including disaster preparation, is taught in elementary school and junior high. Every year, there is a live two-day drill, called Meteor, involving not only emergency response officials, but also large numbers of citizens across the country.[20]

Fifty-five thousand people were evacuated in Cuba in advance of Sandy's landfall there. Fourteen years earlier, when Hurricane Georges was approaching, two hundred thousand were evacuated; military service people and volunteers went door to door to alert people. That storm killed six people in Cuba and at least one hundred in Haiti.

Cuba's discipline is unusual. It results not from wealth, but from a degree of organization that is possible in a country with strong government control over all aspects of life. We may not all want to live in an authoritarian state, but in a disaster, it can work. Haiti, by contrast, is both dirt poor and without strong institutions, and the results are clear.

7

HURRICANE SCIENCE 101

What Is a Hurricane?

Tropical cyclone is the scientific term. Tropical cyclones are called hurricanes in the Atlantic and eastern Pacific. The word *typhoon* is used in the western North Pacific, and in other regions they are just called cyclones. All these mean the same thing.

A cyclone is a storm that has some rotation to its winds, in the same direction as the rotation of the earth: counterclockwise in the Northern Hemisphere, clockwise in the Southern. Because of the Coriolis force, winds blow cyclonically around a center of low pressure. (A rotating circulation in the opposite direction is called an anticyclone, and will have high pressure at its center. A blocking high, like the one south of Greenland in late October, is an example of an anticyclone.)

The adjective *tropical* refers to the low-latitude origin of the storm, but it carries more meaning than that. A mature tropical cyclone has a particular structure, and a particular set of physical mechanisms that make it go. A cyclone that forms at higher latitudes—an extratropical cyclone (sometimes called a winter storm)—has a different structure, and gets its energy from different sources. (*Extratropical* means "outside the tropics"; it does not mean "more than tropical.")

A tropical cyclone, if nothing disrupts it, is circularly symmetric. We are all familiar with the pinwheel appearance of hurricane clouds in satellite images, with bands spiraling out from the eye in the center. The air in the eye is often clear, and warmer than the air outside.

The most intense winds are near the bottom of the eyewall, a towering ring of solid cloud rising into the stratosphere. Farther out from the center, the winds near the surface also circulate around the storm, but they spiral inward, to rise when reaching the eyewall. At the top of the eyewall, the air spirals outward again under the centrifugal force of the rotation. So while the strongest, most destructive part of the wind, under the eyewall, takes a nearly circular path around the storm center, at the same time the storm has what we call a secondary circulation, which goes inward near the surface outside the eyewall, up in the eyewall, and then outward again—in, up, and out.

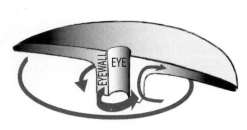

The structure of a tropical cyclone in cutaway view, showing the area filled by cloud with the cloud-free eye at the center. The surface winds spiral in toward the center. The winds circulating around the center are strongest near the surface under the eyewall. Superimposed is the secondary circulation consisting of motion in toward the eye near the surface, up in the eyewall, and out at the top. *(Courtesy of Marit Larson)*

The in-up-and-out circulation drives the strong winds because of angular momentum. Angular momentum is the momentum of something (mass times velocity) on a circular path around an axis, multiplied by its distance from the axis. For a circularly symmetric fluid flow (as in a hurricane), without friction, the angular momentum of any bit of air is conserved, meaning it can't change; the angular momentum is carried along with the air as it moves, unless friction slows it down. If the distance from the axis changes, the velocity has to change inversely, so that the product of the two remains the same. As air near the surface moves closer to the center, its rotation around the center has to speed up. The classic analogy is a figure skater doing a spin: she speeds up when she brings her arms close to her body, slows down when she stretches them outward. The secondary circulation of the hurricane brings the storm's "arms" inward near the surface. Friction of the winds against the surface slows the winds down to some extent—angular momentum is not perfectly conserved—but the "in" portion of the in-up-and-out circulation is what makes the hurricane's winds so powerful.

What Makes a Tropical Cyclone Form?

Tropical cyclones don't appear spontaneously. There is always a weather disturbance of some kind there before. It may be an easterly wave or just a "cloud cluster," a technical term meaning exactly what it sounds like: a bunch of clouds clumped together, visible in satellite images. This predecessor disturbance, being not yet a tropical cyclone, is less organized. The clouds may be popping up in different places and then fizzling out again, rather than staying continuously in one place, and there isn't yet an eyewall or eye, or the very strong winds that go with them. There may be some rotation, and a weak secondary (in-up-and-out) circulation, but these haven't yet gelled. The rotation in the winds is strongest well above the surface, and the surface winds are still relatively weak.

At some point, the system organizes itself. The convection becomes steadier and deeper, and congeals into a circular eyewall. The secondary circulation becomes coherent around a well-defined center, and the rotation strengthens and extends downward, so that strong surface winds begin to spin around the center below the eyewall. If it's a strong hurricane, the eye and eyewall develop to the point that they can be seen clearly from space.

This event is called tropical cyclogenesis, the birth of a tropical cyclone. We don't have a perfect understanding of how this happens, or why it sometimes doesn't happen even though it looks like it might. But we do know a lot about what conditions are necessary for tropical cyclogenesis to occur, and what conditions are likely to prevent it. The following conditions are required of the environment, meaning a large area around the developing storm:

The first ingredient a tropical cyclone needs is a warm ocean surface. The ocean is the source of the tropical cyclone's energy. A rule of thumb historically used by forecasters is that the sea surface temperature has to be higher than around 26.5 or 27 degrees Celsius (about 80 degrees Fahrenheit) in order for a tropical cyclone to form. This rule is sometimes useful as a first guess, but it is sometimes misleading. Sometimes tropical cyclones form over sea surface temperatures cooler than the thresholds. Certainly they sometimes don't form even when there is a predecessor disturbance over sea surface temperatures well over the threshold.

What really matters is not how warm the ocean is, but how warm it is compared to the atmosphere. Because of this, the sea surface temperature threshold rule is not going to be useful at all (not even approximately) as a guide to what will happen under global warming. Although we know that more of the oceans' surface area will have temperatures greater than twenty-seven degrees Celsius in the future than the present, it definitely does not follow that the area in which tropical cyclones form will become larger, or that there will be more tropical cyclones. To the extent that there is a meaningful threshold surface temperature for tropical cyclone formation at all, it will increase as the atmosphere above the sea surface warms, at about the same rate. Understanding how hurricanes will change in the coming warmer climate is, unfortunately, a subtle problem, not amenable to any simple rules that we know.

All that said, as a quick guide *in the present climate*, saying that sea surface temperature has to be at least twenty-seven degrees Celsius (let's just take that as a round-ish number) is not bad as a starting point for outlining the areas of the oceans where tropical cyclones form.

In fact, it's not just temperature at the sea surface that matters. A deeper layer of the ocean has to be warm if the tropical cyclone is to have a good chance of becoming very powerful. As it gets going, the storm's strong winds churn up the ocean beneath it. The turbulence in the ocean causes the water near the surface to be mixed with water from deeper down. Usually the deeper water is colder, so the mixing cools the surface. If the storm doesn't move too quickly, it will still be there when the surface water cools, and the cooler water will weaken it. But this won't happen if the deep water is as warm as the surface, or nearly as warm, so that the mixing doesn't cause much cooling.

The second ingredient necessary for tropical cyclogenesis is high humidity in the atmosphere. The relative humidity, in particular, has to be close to 100 percent, meaning that the amount of water vapor in the air is close to the saturation point, the point at which some of the water vapor must condense. Within the clouds in the pre-hurricane disturbance, the air is already saturated. A cloud is just tiny drops of water (or ice) that have condensed from the vapor; that condensation can't happen at less than 100 percent relative humidity. But for the tropical cyclone to stand a good chance of strengthening, the relative humidity has to be high over a large area, and also up to high altitudes. As the cloud updrafts rise, their turbulence mixes clear air into them from outside. If the clear air is too dry, it will evaporate the cloud droplets and squelch the cloud. In the early stages, before the tropical cyclone is well under way, the clouds are disorganized, and popping up all around a large area, so the whole area needs to be humid enough in order to ensure that the cloud updrafts can keep going strong, giving them a chance to organize.

The third ingredient is that there can't be too much vertical wind shear in the environment around the storm. This term refers to differences in either the direction or the speed of the wind at different altitudes. If the wind is much faster (or slower) at high altitudes than at low altitudes, or if it blows equally strongly but in different directions at high altitudes than at low altitudes, we say there is strong shear. If we were to put a gigantic paddle wheel in the atmosphere, with its axis in the horizontal so it were to sit like a Ferris wheel, wind shear would make the wheel turn by pushing harder on the top than the bottom (or vice versa).

Strong wind shear can tear the tropical cyclone circulation apart before it has a chance to form. The tropical cyclone vortex naturally wants to stand straight up; vertical shear tries to tilt it over, as the high-altitude winds blow the top one way and the low-altitude winds blow it another (or the same way at different speeds).

Once the storm is mature, it is resistant to being torn apart in this way. But by tilting and distorting the updrafts in the eyewall and in the rainbands (the spiral arms of cloud often visible in satellite images of hurricanes), shear allows a little more dry air into the clouds. That evaporates some cloud water. The evaporation cools the air, making it heavy enough to sink in downdrafts. The cool air from the downdrafts mixes into the warm, moist surface air that is about to ascend in updrafts, making it less buoyant and slowing down its rise. Shear is like an immune deficiency, weakening the hurricane's defenses against infection by dry air.[21]

The fourth and last key ingredient for hurricane formation is rotation. In order for the storm to generate its strong rotating winds, there has to be some rotation there to work with at the start. The inward flow at the bottom of the storm brings in high angular momentum and increases the spin as the radius decreases, just like the skater bringing her arms in as she spins. The skater has to be spinning already, though. If she's standing still, bringing her arms in won't make her spin. It's the same with the incipient hurricane.

In almost all cases, the initial rotation comes from the earth itself. Because the atmosphere is stuck to the spinning earth by gravity, it has angular momentum about the earth's axis—kilogram for kilogram, the same amount as solid earth or water. But the storm needs angular momentum not about the earth's axis of rotation, but about the vertical, an axis sticking straight up out of the surface, wherever the storm is.

The closer the vertical axis is to the axis of rotation, the more angular momentum the storm inherits from the planet. At the pole, the earth's axis and the vertical are the same. At the equator, they are perpendicular, and the earth's rotation does nothing for an incipient hurricane. It is nearly impossible for a tropical cyclone to form at the equator, even if all the other ingredients are there. The sea surface temperature can be thirty degrees Celsius, the atmosphere can be saturated with water vapor, the wind shear can be zero, and a vigorous cloud cluster can be churning away—but because there's no rotation, nothing will happen. If we look at a historical map of the locations of tropical cyclone formation, there is a conspicuous gap a few degrees of latitude wide around the equator. Just that few degrees (a couple hundred miles from the equator) is enough to make the difference, but any closer than that to the equator and you're pretty safe from hurricanes.

Besides all the environmental factors, there is an element of chance in whether any particular cloud cluster or easterly wave manages to become a tropical cyclone. Forecasters watch the satellite images, and any other data they may have (from buoys, ships, surface stations if the system passes over land, aircraft observations if they're really lucky) from hour to hour, trying to assess how the pieces are coming together. Are the cloud tops showing the convection getting deeper and more organized? Is the surface pressure dropping? Are the winds increasing? How are the large-scale conditions changing: Is the storm moving into an area of lower wind shear or higher sea surface temperature? Or, if the storm is not moving rapidly, are any of the key conditions changing due to the evolution of the atmosphere itself on the large scale? Is an MJO active phase, for example, setting in? And in recent years, it has become useful to ask: what do the models say is going to happen?

Our computer weather models have become quite skillful at predicting tropical cyclogenesis. They're often still wrong about it—they can fail to predict when a storm will form, or will predict one where none does—but they're right much more often than they were in the past, and often enough to be quite useful for forecasting. This is particularly true when the forecaster looks at the whole set of models being run around the world. If most of the models start to agree that a hurricane is going to form, it's a good bet that it will. This wasn't true until recently; ten years ago, the forecasters didn't put a lot of stock in the models when trying to decide if a hurricane would form. But they do now.

This tells us that while there is some randomness and unpredictability involved in determining whether those incoherent clouds will start to swirl, it isn't quite as much as we might think.

The detailed structure of the pre-hurricane disturbance is never exactly right in the models. For one thing, the models don't have the resolution. Each model has a grid spacing that plays the role of the size of a pixel in a digital camera image. You can't make out an object in the picture unless it's a lot bigger than a single pixel, so that different pixels of different brightness and color can be used to construct the image's different parts. It is the same with a computer model of a storm: the storm has to be much bigger than the pixel.

Today's weather prediction models have grid spacings in the range of ten or maybe twenty kilometers (six to twelve miles). That's fine enough to make out the gross structure of an easterly wave, say, but not all the different cloud updrafts, downdrafts, and smaller circulations connected to those inside the wave. Those features may make a difference in exactly how the storm evolves, but at least much of the time, their details must not matter. If they did, the models, with their oversize pixels and fuzzy impressionistic renderings, wouldn't be able to predict tropical cyclogenesis happening in the right places at the right times as often as they do.

The models had gotten Sandy's formation right. They had disagreed among themselves, critically, on how Sandy's trajectory would end. At midweek, that disagreement had not yet been resolved. But it was becoming clear that the worst-case scenarios were not just the fantasies of weather nerds. The storm's transformations were bringing those scenarios closer to reality.

8

WEDNESDAY, OCTOBER 24

On the night of Wednesday, October 24, Sandy had just left Jamaica. Landfall in Cuba was imminent. The model guidance also indicated quite clearly that the Bahamas would be next. In the NHC discussion of 11:00 p.m. EDT, forecaster Jack Beven wrote:

THE HURRICANE IS CURRENTLY BETWEEN A MID/UPPER-LEVEL LOW OVER THE NORTHWESTERN CARIBBEAN AND A MID/ UPPER-LEVEL RIDGE OVER THE EASTERN CARIBBEAN. THE DYNAMICAL MODELS FORECAST SANDY TO MOVE AROUND AND EVENTUALLY MERGE WITH THE LOW . . . WHICH SHOULD PRODUCE A NORTHWARD MOTION FOR 24 HR OR SO FOLLOWED BY A NORTH-NORTHWESTWARD TURN.

In the Northern Hemisphere, the wind turns counterclockwise around a low and clockwise around a high, or "ridge." So if you have a low to your west and a high to your east, as Sandy did, the winds from both are blowing from south to north. Hurricanes are steered along their tracks mainly by the winds in the larger environment around them. Hence the confident prediction that Sandy would be steered northward in the short term, into the Bahamas after Cuba.

Beven went on:

FROM 48–96 HR . . . A DEEP-LAYER TROUGH MOVING INTO THE EASTERN UNITED STATES SHOULD STEER THE MERGED SYSTEM GENERALLY NORTHEASTWARD. THE GUIDANCE BECOMES DIVERGENT AFTER 96 HR . . . WITH THE ECMWF/GFDL/NOGAPS TURNING THE CYCLONE NORTHWESTWARD . . . AND THE GFS/ HWRF SHOWING A NORTHEASTWARD MOTION.

"Guidance becomes divergent" is forecaster language for "the different computer models we are looking at are not telling us the same thing." The models from the European Centre for Medium-Range Weather Forecasts (ECMWF, in Reading, UK), the NOAA Geophysical Fluid Dynamics Laboratory (GFDL, in Princeton, New Jersey), and the U.S. Navy's Operational Global Atmospheric Prediction System (NOGAPS, in Monterey, California) were predicting that the model would turn westward in four days—late on October 28 or early on the twenty-ninth. The Global Forecast System (GFS, run by NOAA's National Center for Environmental Prediction, in Maryland) and the Hurricane Weather Research and Forecast (HWRF) model, on the other hand, were taking it out to sea.

Those were the "deterministic" model runs—single runs of each model, using the highest possible resolution. In addition, a lower-resolution version of each model is run many times, to produce an "ensemble," with the different ensemble members giving slightly different results. These larger ensembles showed the same schism as the deterministic runs. Within the GFS ensemble, for example, some members made landfall, while more went out to sea. Within the ECMWF ensemble, more made landfall, while fewer (but still some) went out to sea.

When different models disagree, forecasters cannot be sure which is right. This is particularly so four or five days in advance of a storm. The longest lead at which NHC issues a forecast is one hundred twenty hours— five days. That was just around the time of the split in the models; and at this time, the forecast was moving Sandy along its future track a little more slowly than it actually would move. So at the hundred-twenty-hour lead time, NHC did not yet have to commit fully to the left or right turn.

People watching the news in the mid-Atlantic and northeastern United

States were beginning to become aware that a storm was heading north, and that the left turn was possible. Sandy chatter had already been in the weather blogs for a couple of days, and it now was moving into the mainstream media. For the most part, the stories reflected the uncertainty fairly enough. Landfall in the United States was a possibility, not a certainty.

NCEP   **ECMWF**

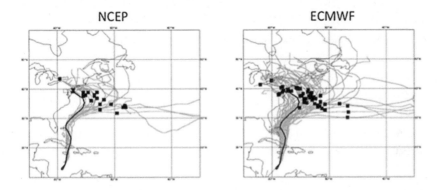

"Spaghetti" plot showing forecast tracks of Sandy from different members of the GFS model ensemble, run at the National Centers for Environmental Prediction (NCEP) in the United States and from the European Centre for Medium-Range Weather Forecasts (ECMWF) model ensembles, on October 24 at 00 UTC. The squares indicate the position of the forecast storm on the date when landfall actually occurred. Compared to the ECMWF model, a greater fraction of the NCEP ensemble members' storms was heading out to sea rather than making landfall. *(© 2012 ECMWF, image courtesy Linus Magnusson)*

On Wednesday, the models were still split. Most of the European Centre's members were making the left turn. Many members of the United States' GFS model ensemble were still going out to sea, but more were starting to defect to the European solution.

With much greater uniformity than they predicted landfall, the models were predicting that Sandy would, over the succeeding several days, combine with an extratropical, or winter, storm. The media coverage began to include this prediction. Being a hybrid of two different entities, this predicted combination of two storms reminded someone of Frankenstein's monster, a hybrid person stitched together from pieces of different bodies in Mary Shelley's novel. Sandy was given the name Frankenstorm in many of the first media stories. Perhaps the fact that the storm would arrive close to Halloween helped this name to stick.

In his discussion on the night of Wednesday the twenty-fourth, Beven wrote:

GIVEN THE CURRENT INTENSIFICATION . . . THE NEW INTENSITY FORECAST NOW CALLS FOR SANDY TO REMAIN A HURRICANE WHILE CROSSING CUBA. WHEN THE CENTER EMERGES NORTH OF CUBA . . . IT IS FORECAST TO INTERACT WITH AN UPPER-LEVEL TROUGH THAT WILL PROVIDE STRONG UPPER-LEVEL DIVERGENCE AND BAROCLINIC ENERGY. WHILE THIS WILL ALLOW SANDY TO STAY AT OR NEAR HURRICANE STRENGTH . . . IT IS LIKELY THAT THE WIND FIELD WILL SPREAD OUT AND THAT THE CYCLONE WILL LOSE SOME OF ITS TROPICAL CHARACTER IN THE PROCESS.

The "upper trough" was an extratropical disturbance, a distortion of the jet stream, dipping to the southeast from high above the North American continent. When an extratropical storm intensifies, such a disturbance typically provides "upper-level divergence and baroclinic energy" to strengthen an incipient surface low-pressure system. While hurricanes get their energy from the ocean below, extratropical storms draw energy from the jet stream, high up in the atmosphere. Beven's technical language meant that Sandy was about to get a jolt of extratropical energy from above.

With that energy would come structural changes. Sandy would take on some extratropical characteristics. Among those, the wind field would "spread out." Extratropical storms are usually not as powerful as hurricanes, but they are often much larger. Sandy's partial hybridization over the Bahamas would make it a much larger storm. The extent of this expansion would become clear soon. The implications for those in its path would be as huge as the storm itself. To understand what was happening to Sandy, we need to know more about extratropical storms.

9

WINTER WEATHER

An extratropical storm is the kind of storm you have experienced many, many times if you are an adult and have lived any significant part of your life outside the tropics. They are typically strongest in winter, and perhaps for that reason are sometimes called winter storms. I don't like that name, because they occur year round. I will use the scientific name extratropical storm or extratropical cyclone. Whether your home experiences a real winter (one where it gets cold enough that you need a jacket even on a sunny day) is, however, a good indication of whether you live in the extratropics, which is scientific for "any place that is not in the tropics." In terms of lines on the map, that's anywhere more than thirty degrees of latitude or so away from the equator. There isn't a precise boundary, though; extratropical storms can dig down into the tropics sometimes.

Even though such storms occur throughout the year, the correspondence between where there is winter and where there are extratropical storms is not coincidental. It's always warm in the tropics. The temperature changes a little with the seasons, and the seasons are defined more by rainy versus dry than by cold versus warm. It pretty much never gets cold in the tropics, except at the tops of mountains. As one moves closer to the pole, though, particularly in the winter hemisphere, the sun gets lower in the sky and the temperature drops.

Climatologists say there is a "temperature gradient" between the equator and the pole. But that gradient doesn't really start to steepen until we get out of the tropics. If we make a map with contours of constant temperature (just like a topographic map, hence my use of the word *steepen* by analogy), the contours become closer together as one gets into the midlatitudes. There is little difference in temperature between the typical place on the equator and the typical place at, say, fifteen degrees north. Montreal, on the other hand, is less than fifteen degrees north of Dallas; the winters in the two places are so different that they might be on different planets.

The temperature gradient is the essential ingredient in an extratropical storm. You experience it directly when you're in such a storm. Almost always, the storm is associated with one or more fronts, zones of particularly sharp temperature contrast. Often, when a storm first arrives the wind is from the south—we're in the Northern Hemisphere now; Southern Hemisphere readers, forgive me—and it is relatively warm. The air may also be humid; the heaviest precipitation, whether rain or snow, typically falls during this phase. Then, very abruptly, the wind shifts to blow from the north, and the temperature dives. The skies often clear shortly, but the cold wind stays.

The presence of fronts tells you right away that an extratropical cyclone is a different animal than a tropical one. The thing that stands out most about tropical cyclones in satellite images is their circular symmetry: the pinwheel of high cloud with the eye in the center. That is a direct indication that the storm is also circularly symmetric beneath the cloud. The characteristics of the storm will change as one moves away from the eye, but it doesn't matter whether one moves away toward the north, south, east, or west. By contrast, an extratropical cyclone's fronts arc out from its center, dividing the pie into cold and warm sectors. You can tell easily which sector you're in by the weather you experience on the ground.

The asymmetry is visible from space. The typical extratropical cyclone's clouds show a comma shape, rather than a symmetric spiral, in satellite images. Clouds stream northward from a point near the storm center (which itself isn't nearly as easy to identify from space as a hurricane's eye is), then expands into a thick band as it wraps around to the east and then south.

The boundary between the southward-dipping band and the clear air to its west usually marks the cold front, behind which dry air from higher latitudes is descending as it spirals in toward the pressure minimum in the center (none of that, other than the cloud, being visible in the image). Just on the east side of the cloud band, warm air from lower latitudes is moving poleward and rising.

The rising air moving north in the cloud band expands and cools as its pressure drops with altitude. The cooling causes water vapor in the air to condense; hence the cloud, and the precipitation under it. As this moist, warm air moves north, it also spirals to the left; it meets the cold air in a warm front, and rises over it. Rather than the symmetric in-up-and-out circulation of a hurricane, the extratropical cyclone's circulation, like everything else about the storm, is asymmetric. Distinct air masses come in from different directions, meet in fronts, and slide under and over each other.

The fronts, clouds, and low pressure are what we experience most directly at the surface, but much of the action in an extratropical cyclone occurs much higher up, at the level of the jet stream. The jet stream is a zone of very strong wind high above the earth. The core is around seven or eight miles up, roughly where jet planes fly. (No relation: a jet is just a narrow, fast current in a gas or liquid; the jet in the jet stream is in the atmosphere, while that in a jet plane is in the engines.) The jet stream blows from west to east, more or less, around the planet. But it undulates, and those undulations are the extratropical cyclones.

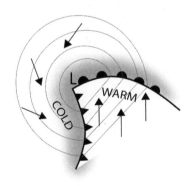

Schematic surface weather map showing the structure of a developing extratropical cyclone. The contours are isobars (contours of constant surface pressure), with low pressure in the center. The warm front is marked by the semicircles, the cold front by the triangles; the darker shading along the fronts indicates regions of precipitation; and the arrows indicate the surface winds. *(Courtesy of Marit Larson)*

Just as the extratropical cyclones are inherently connected to the north–south temperature gradient, so is the jet stream. The jet stream sits atop the contrast between warm tropical and cold polar air; it rides quickly along that temperature drop, rather than across it, because of the Coriolis force, a consequence of the rotation of the earth. If it could, the jet stream would blow straight west to east, but it can't. It couldn't do so even if there were no mountains, or continents, or anything else underneath to disrupt its flow. The extratropical storms are there, in fact, because the jet stream is inherently unstable.

The instability of the jet stream and the extratropical weather systems that result from it are both the subject of a tremendous body of modern science. Dynamical meteorology and geophysical fluid dynamics[22] are founded strongly in observations of real weather phenomena, complemented by systematic explanations of those phenomena using the methods of theoretical physics and applied mathematics, all developed to a high level of sophistication over the course of the twentieth century.

Let's accept, for the sake of argument, that the jet stream is related to the pole-to-equator temperature contrast, that the whole system is inherently unstable, and that the result of the instability is extratropical cyclones. Then there are a few more obvious differences between those extratropical cyclones and their tropical counterparts that we can explain as logical consequences of those facts.

1. Tropical cyclones form over ocean and die quickly over land. Extratropical cyclones, on the other hand, do just fine over land. They don't get their energy from the ocean, but from the temperature contrast, which can be as strong over land as ocean. (Residents of the northeastern United States or East Asian coasts may know that storms often strengthen fast right at the coast in winter; that's because the land is cold while the ocean is relatively warm, setting up a particularly strong temperature contrast for the storms to feed on.)

2. The strongest winds in a tropical cyclone are right at the surface, where the secondary (in-up-and-out) circulation is squeezing angular momentum in toward the center; the strongest winds in an extratropical cyclone are high up, since that's where the jet stream is. (The extratropical cyclone does, however, spin up a secondary circulation. It's also in, up, and out, just much more asymmetrically so, with the upward-moving air coming in from the lower latitudes while colder air from high latitudes sinks below it.)

3. A tropical cyclone always forms from a precursor disturbance; it can't appear out of nowhere. An extratropical cyclone can, because it's growing from an inherently unstable system. Think of a ball sitting at the top of a hill with a rounded peak. That's an unstable system. If pushed even slightly to either side, it will fall. The tropical cyclone does not result from such a simple instability; it needs a good push, in the form of an easterly wave or a cloud cluster, to get going, and once it does, it becomes something else entirely from what it was. (If we want to push the analogy a little too far, it's a situation more similar to a hill with a pit dug in the top of it, and the ball is sitting in the pit; it won't roll downhill by itself, but if it gets a big enough kick, it will.)

These distinctions are useful and necessary. Classification is an essential part of science. Meteorology, in this, is no different from zoology. We first classify storms, like animals, by their outward characteristics. In the case of animals, these include their physical traits: size, shape, those things you could tell from a dead specimen. (In the case of storms, this would be a single satellite image or weather map.) Then, if you have access to live specimens in their habitats, you can have more dynamic information about each species: its habit, life cycle, what it eats and what eats it, the environments in which it lives.

A more mature scientific field will go beyond classification to explain the reasons things are as they are. Distinctions between species should not stop at appearance. Animals of the same species can mate and produce offspring, whose traits are passed along in DNA. Different species may be related to one another by sharing common ancestors, close or far away on the evolutionary tree. Their changes over time result from random mutations, reinforced by their interactions with their environment through natural selection. Similarly with storms. Tropical and extratropical ones come from different progenitors, in different environments; derive their energy from different sources; and thrive (or wither) under different conditions.

But the atmosphere has no clear boundaries. The distinction between a storm and its environment, or between two storms, is not sharp. And unlike an animal's, a storm's species is not fixed at birth, but can change. Occasionally, the individual that comes out of the change can defy the expectations that rest with the classification system, as well founded and scientifically mature as that system is. Some animals are not quite fish or fowl.

10

THURSDAY, OCTOBER 25

The Transition Begins

At 11:00 a.m. EDT on Thursday, October 25, Sandy was leaving the north shore of eastern Cuba. The passage over the island had weakened the storm slightly. The satellite images showed a ragged appearance. NHC forecaster Michael Brennan wrote:

THE SATELLITE PRESENTATION OF SANDY HAS BECOME A LITTLE LESS IMPRESSIVE AFTER MOVING OVER EASTERN CUBA. THE EYE IS NO LONGER APPARENT IN IMAGERY OR IN OBSERVATIONS FROM THE AIRCRAFT . . . AND THE CDO HAS BECOME SOMEWHAT DISRUPTED ON THE WESTERN SIDE BY SOUTHWESTERLY FLOW FROM AN UPPER-LEVEL LOW TO THE WEST. HOWEVER . . . AIRCRAFT DATA INDICATE THAT THE CIRCULATION ASSOCIATED WITH SANDY REMAINS QUITE STRONG . . . WITH A PEAK FLIGHT-LEVEL WIND AT 700 MB OF 126 KT AND A CENTRAL PRESSURE OF 964 MB. DATA FROM THE SFMR AND DROPSONDES SUGGEST THAT THE WINDS FROM FLIGHT LEVEL ARE NOT CURRENTLY MIXING DOWN TO THE SURFACE AT THE TYPICAL RATIO . . . SO A BLEND OF THE DATA SUPPORT MAINTAINING AN INTENSITY OF 90 KT FOR THIS ADVISORY.

CDO stands for "central dense overcast," meaning the area of thick cloud around the center of the storm. The shape of the CDO is one of the primary features used in the Dvorak technique, the standard method for determining the strength of a storm from satellite images. Named for its inventor, Vernon Dvorak, who published it in 1984, the method consists of a systematic, step-by-step set of pattern recognition tasks laid out in a flow chart. All hurricane forecasters are trained in it, because satellite images are sometimes all forecasters have to go by when trying to estimate a tropical cyclone's intensity. The more circularly symmetric the CDO, all else being equal, the stronger the storm. So the disruption cited by Brennan indicated weakening.

In this case it wasn't necessary to rely solely on satellite images, because the Hurricane Hunters were flying the storm. At flight level, corresponding to 700 millibars air pressure, or about two miles above the surface, the winds were directly measured at 126 knots. Winds at flight level are usually stronger than at the surface. The surface winds are slowed down by the drag exerted by the ocean; even more so because the disturbed seas, churned up by the winds themselves, present a rough surface. The winds aloft can't be too radically different from the surface winds, though, because the turbulence in the air, including the updrafts and downdrafts in the eyewall, mixes together the momentum of the air at different heights. That's the "mixing down" the discussion mentioned; this speeds up the surface winds and slows down the winds higher up. The "typical ratio" of surface winds to flight-level winds is about 0.9, meaning surface winds are 90 percent of those at flight level.

The 90 percent figure comes from analysis of many dropsondes over time.[23] A dropsonde is essentially the same set of instruments as on a normal weather balloon, but instead of being launched upward on a balloon, it's dropped down out of a plane. It directly measures temperature, pressure, and humidity, and winds are measured by tracking the sonde's motion very precisely by GPS. It sends its data back by radio, then falls into the sea.

In this case, the dropsondes agreed with the SFMR's microwave scans of the sea surface that the ratio was only about 70 percent, rather than 90 percent—90 knots maximum sustained surface winds, compared to 126 at flight level. Ninety knots is still high in the category two range—no joke as Sandy headed north—but the storm would continue to weaken, as anticipated by the forecast.

The first paragraph of Brennan's discussion continued:

A CONTINUED INCREASE IN SHEAR AND INTERACTION WITH
THE UPPER-LEVEL LOW SHOULD RESULT IN GRADUAL WEAKEN-
ING DURING THE NEXT COUPLE OF DAYS. HOWEVER . . . DURING
THIS TIME THE OUTER WIND FIELD OF SANDY IS EXPECTED TO
EXPAND . . . AND SANDY IS FORECAST TO BE A LARGE CYCLONE
AT OR NEAR HURRICANE INTENSITY THROUGH THE END OF THE
FORECAST PERIOD.

This was good news for the Bahamas, but ominous for the United States
of America.

It was a mixed message because Sandy, though still in the tropics (at
22.4 N, 75.5 W), was starting to morph into a hybrid of two different
storms. The discussion cites a "continued increase in shear and interaction with the upper-level low" as factors causing weakening. Shear is bad
for tropical cyclones, and good for those who are in their path. (Meteorologists' language often implies value judgments from the storms' point
of view, a perpetual problem when communicating with people whose
priorities are in conflict with those of a hurricane.) By tilting the storm
over, shear gives dry air better odds of getting into the eyewall, where it
can evaporate cloud drops, cooling the air and reducing the buoyancy of
the updrafts.

The upper low, though, was a separate system—not a product of
Sandy, but an external actor, moving in from the west. Though centered
at twenty-five degrees north—far to the south of the jet stream, which at
this moment was bulging north over eastern Canada, as high pressure
and calm weather held over the eastern United States—this upper low had
some characteristics of an extratropical system. The complex, tricky thing
is that extratropical systems have the *opposite* response to shear that tropical cyclones do.

Extratropical storms like shear. They thrive on it; they wouldn't exist
without it.

Early in the life of an extratropical storm, when it is growing in strength,
it is tilted westward with height. In other words, if we have a low-pressure
system at the surface, with winds circulating cyclonically around it (in the
same direction as the earth's rotation; counterclockwise in the Northern
Hemisphere, clockwise in the Southern), we also have a low-pressure center

with cyclonic winds at high altitude, around the level of the jet stream, but the upper center is west of the lower center.

The jet stream blows from west to east, and is stronger at high altitudes than low; we say it has "westerly shear," since winds from the west are called westerly, and we name the shear according to whatever direction is stronger higher up. This means the low-pressure system as a whole, by being tilted westward with height, is tilted *against* the shear. If we draw a line connecting the lower-level and upper-level low-pressure centers, and imagine the jet stream just blowing that line along with it, the top would move to the east faster than the bottom, and the line would want to move to a more upright position.

Your intuition might be to assume that an upright storm is more powerful than a tilted storm. If so, you would be right. As the shear pushes the upper low over the surface low, the whole system strengthens. When we put a cyclonic low-pressure system at high altitudes on top of another one at the surface, we strengthen the surface one.

The angular momentum part of the story also has to work out. If the cyclonic winds are spinning up, there must be a secondary circulation pushing air inward toward the center, near the surface (pulling in the skater's arms); that air has to go somewhere once it gets to the center, and it can't go down, so it goes up, and then out at high levels. The tilting of the vortex to a more upright position by the shear implies an in-up-and-out circulation something like one in a hurricane. Since the whole thing is driven by the relative motion of the upper system over the lower one, we can think of the upper system sucking a column of air upward under it as it moves in the shear. The rising air is pushed outward as it reaches the top of the storm. This outflow is the "upper-level divergence" that forecasters talk about. It's an indicator of an upper low acting to energize a system below it as vertical shear brings the two into alignment.

The upward suction induced by the upper low occurs over an area much larger than a typical hurricane eyewall. The contraction of angular momentum and associated spin-up of cyclonic wind at the surface is similarly large. This makes extratropical cyclones less intense, because they don't draw angular momentum as tightly in toward the center, but it also gives them much larger footprints compared to tropical cyclones.

When we have a hurricane, vertical shear, and an upper low to the west, it isn't obvious what the result will be. The shear brings in dry air that disrupts the convection, and that weakens the hurricane. Some of that shear, on October 25, was due to the upper low itself. The low was strongest at upper levels; as the hurricane was brought into that circulation on the low's eastern flank, it felt the shear of the upper cyclonic circulation against the weaker flow below.

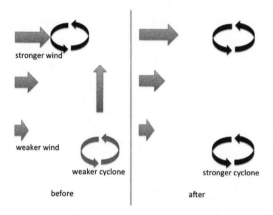

Schematic diagram showing the interaction of vertical shear with a developing baroclinic cyclone. The arrows to the left indicate the mean winds associated with the jet stream, stronger at high altitude than at low. The circular arrows represent the upper- and lower-level circulations associated with the storm. In the first frame *(left)*, the system is tilted, with the upper cyclone to the west of the lower one. Because of the stronger winds aloft, the upper cyclone moves eastward more rapidly than the lower one. This brings the two systems into vertical alignment as shown in the second frame *(right)*, which occurs later in time. The alignment strengthens the surface system. This process is accompanied by large-scale ascent, indicated in the first frame.

Some of the shear, though, was in the broader environment, bringing the upper low in closer to the hurricane. As shear in the jet stream brings an upper low closer to a surface low, it strengthens a normal extratropical storm. Here the same thing was happening, but with Hurricane Sandy. The shear was forecast to weaken the preexisting hurricane but strengthen this new, hybrid system that would begin to emerge on top of it. The hybrid might not have the potential to have winds as strong as Sandy's had been over Cuba, but it could keep them at category one, hurricane strength. At the same time, the strong winds would cover a much larger area, typical of an extratropical system. Sandy would keep enough of a warm core, and enough convection in the center, to retain the title of hurricane. But it had already begun to change, down in the Bahamas.

The Track Forecast

The second paragraph of Brennan's 11:00 a.m. discussion on Thursday the twenty-fifth addressed the forecast track of Sandy. Since the day before, that forecast had changed.

Track forecasts always change from one day to the next. If nothing else, because NHC issues forecasts as much as five days in advance, each forecast sees one day further than the one made the same time on the previous day.[24] Besides that, though, there is new information. The storm evolves, new data come in, and the models are run again. The first four days of the forecast, though they were also in yesterday's forecast (as the second through fifth days), can also change, as they grow closer in time.

Up to Wednesday, there had been a major disagreement between the different models on the late portion of the track. The ECMWF, since days earlier, had been confidently predicting a left turn, into the northeastern United States. Some other models had agreed. The GFS (the main American model) had been recalcitrant, taking Sandy out to sea. By Thursday, the GFS blinked. The entire multimodel ensemble had locked in on the left turn. The job of forecasting the track became simpler as the implications of the forecast became much graver.

After a couple of introductory sentences, the second paragraph of Brennan's discussion explained the storm's narrowing options:

SANDY SHOULD MOVE NORTHWARD TODAY AND SLOW DOWN AS IT TURNS NORTH-NORTHWESTWARD ON FRIDAY WHEN IT INTERACTS WITH THE AFOREMENTIONED UPPER-LEVEL LOW. SANDY SHOULD THEN TURN NORTH-NORTHEASTWARD BY 72 HOURS AHEAD OF A LONGWAVE TROUGH MOVING THROUGH THE EASTERN UNITED STATES. AFTER THAT TIME . . . THE TRACK MODEL GUIDANCE CONTINUES TO TREND TOWARD A NORTH-WESTWARD TURN AS SANDY INTERACTS WITH AN AMPLIFYING SHORTWAVE TROUGH OVER THE MID-ATLANTIC STATES.

This "shortwave trough" was a new upper-level low-pressure system, entirely separate from the one already causing Sandy to change. The *short* in "shortwave" refers to the relatively small size of the system. This is distinct from the much broader southward dip in the jet stream, associated with the negative NAO, bringing cold air behind it from Canada and into the United States, from the Rockies well to the east. Those large troughs, which we might call "longwave," can sit in place for days or even weeks, while shortwaves swing more rapidly through them. This shortwave was already causing snow in Colorado. The models had it moving east as Sandy moved north. When the two storms got close enough, they were forecast to merge. This second, predicted hybridization, stronger than the first already under way, was the reason for the nicknames Frankenstorm and Superstorm. But the interaction would also cause the left turn, bringing the hybrid storm onshore.

Up until the day before, the forecast had implied that a catastrophe was possible for the U.S. East Coast. Now that catastrophe had become almost certain.

This change was not, however, accompanied by any dramatic change in media discussions or government preparations. Perhaps this was because, while it was now becoming certain that the storm would make landfall, exactly when or where was still unclear. The NHC discussion of October 25 went on to characterize the uncertainty still remaining in the track forecast:

HOWEVER . . . THERE REMAIN SIGNIFICANT DETAIL DIFFERENCES IN REGARD TO THE TIMING OF THIS INTERACTION AND WHERE THE NORTHWEST TURN WILL OCCUR. THE ECMWF AND GFDL MODELS SHOW A QUICKER TURN . . . BRINGING THE CENTER INLAND BY DAY 5. MUCH OF THE REST OF THE GUIDANCE SUGGESTS A MORE GRADUAL TURN WITH THE CENTER STILL OFFSHORE AT 120 HOURS. THE NHC FORECAST HAS BEEN SHIFTED WESTWARD AT DAYS 4 AND 5 AND LIES ABOUT HALFWAY BETWEEN THE ECMWF AND THE GFS ENSEMBLE MEAN AND CLOSE TO THE ECMWF ENSEMBLE MEAN.

In other words, NHC was making a call, because that's its job; but the fact was, while it was now clear that someone was going to get hit, it was still too early to say who. But Brennan reiterated that the chance that the whole Eastern Seaboard would make it into the next week without a disaster of some kind was quickly evaporating:

REGARDLESS OF THE EXACT TRACK OF SANDY . . . IT IS LIKELY THAT SIGNIFICANT IMPACTS WILL BE FELT OVER PORTIONS OF THE U.S. EAST COAST THROUGH THE WEEKEND AND INTO EARLY NEXT WEEK.

"It is likely that significant impacts will be felt" is deadpan, dispassionate scientist language. It means "this is no joke."

Seeing the Future; the Left Turn

Sandy at this point already had been in the media for a few days. The early coverage had reflected both the hybrid nature of the forecast storm and forecasters' uncertainty regarding its track. We might get it, we might not; different models disagreed. Now that the models were beginning to lock in a consensus that there would be landfall on the Eastern Seaboard, the media coverage began to change.[25] Stories on Thursday and Friday focused more intently on the hybrid nature of the storm, using the "Franken" and "Super" prefixes.[26] The possibility of a "worst-case scenario" for New York City, with severe flooding if the landfall point were to the south, began to be taken more seriously, though it was nowhere near certain yet.[27] Though many forecasters and scientists emphasized the potential for an event that could be different from, and very possibly worse than, anything in the living memories of those in the region, they tended to emphasize the merging of the hurricane and the extratropical trough. They compared Sandy to the "perfect storm" of 1991, a storm that did relatively little harm in the United States but that had been made famous by Sebastian Junger's book *The Perfect Storm*, and the film based on it starring George Clooney.

The media reports, each featuring quotes from forecasters or scientists from academia or the private sector, didn't focus particularly on the fact that Sandy was being forecast to approach the East Coast on a westward track. Even NHC's discussions, though they explained in detail the factors influencing the track, including the left turn, didn't comment on it beyond that. But that westward track was the most unprecedented aspect of the forecast.

Hybrid storms are not all that unusual; a significant fraction of hurricanes undergoes extratropical transition (the process of changing from a symmetric, warm-cored tropical cyclone to an asymmetric, cold-cored extratropical one) as they move to higher latitudes. Most of the time, this doesn't make the news. Often it happens over the ocean, and the storms never come back; the same jet stream that brings the shear to cause the transition blows them eastward, away from land. Those few that do make landfall often weaken as they make the transition, leaving hybrid storms less dangerous than the hurricanes they replaced.

Sandy was forecast to be very large, and analysis afterward would judge it the largest Atlantic hurricane on record. But we have only around thirty years of good data on the sizes of hurricanes. By contrast, we have more than one hundred sixty years of relatively good data on hurricane tracks in the Atlantic. In that time there has never been one that has come into the northeastern United States on a westward track like Sandy's while maintaining hurricane-strength winds. Not even close.

Human beings function efficiently by assuming, largely unconsciously, that the situations we face now are similar to those we have faced before.[28] When something truly unprecedented happens, we usually don't see it coming. If it's something bad that affects a lot of people, we call it a "black swan." The financial crisis that began in 2008 was an example. Some professionals did predict it beforehand, but most didn't, and even those who did couldn't say exactly when it would happen. Certainly governments around the world weren't well prepared for it.

Here, in Sandy, we had a truly unprecedented event impending. Through interactions with two extratropical storms, a hurricane would reach the largest size ever observed, while simultaneously intensifying and making a left turn like nothing in over a century and a half of recorded hurricane history. It was all forecast very clearly. The whole scenario was known to be a possibility over a week ahead, and by several days ahead it was becoming pretty clear that it was likely. There was no serious debate, as the weekend drew closer; no analogy to the bears saying the housing market would tank and the bulls saying the good times would continue.

Weather prediction is different because we don't have to go by experience. We don't just know what inputs have led to what outputs in the past; we can see the inside of the black box. Weather follows the laws of physics, and we know those.

But knowing the rules is just the beginning. The laws of physics are simple, but working out their consequences in the atmosphere is an extremely complex task. To do it, we need a big infrastructure.

We need big, powerful computers; sophisticated and complex software encapsulating the known laws of physics as applied to the atmosphere, in the form of numerical weather prediction models; a global system of observations, including weather balloons, satellites, surface measurements, radars, and the rest, to tell the models what the weather is right now, as a starting point. Most of all, we need scientists who understand how it all works.

We have all this. It is the result of many decades of research. The work was done by scientists in many countries, but it started with a few scientists in Scandinavia. To appreciate the forecasts of Sandy for the achievement they were, we need to learn some history.

11

WHERE WEATHER FORECASTS CAME FROM

Career Changes: Vilhelm Bjerknes, Turning Weather into Physics

I wanted to be a musician when I was young. In high school and college, and for a few years after, I played the trumpet, and some other instruments less seriously. I was into jazz, composition, and music theory. I studied both music and physics at Wesleyan University, in Middletown, Connecticut. I was a more serious student of music. I got a lot of Bs in physics, some of them given charitably.

When I got out of college, I went back to New York City to try my hand at being a professional musician. I didn't get far. It was clear pretty quickly that I wouldn't be able to pay rent as a trumpet player. I was lucky to get a job working as a sound engineer in a studio that did postproduction for TV and radio commercials. I learned a lot there, but I lost interest after a year or so. When Marit, my girlfriend at the time—she's my wife now—moved from New York to Berkeley, California, to go to graduate school, I followed her there. During our two years there, I decided to go back to school myself, to study physics.

If I were to get a PhD in physics, I knew I would have to choose a specialty within that field. Having been out in the world a little, I understood that I should choose carefully, as the choice would influence my career prospects afterward. Marit, a passionate environmentalist since her childhood, argued to me that meteorology would be a better choice than physics. Global warming was an important problem, she said—this was 1992, and she was one of the relatively few people who took it seriously back then—and physics was the right background for it. Initially, I wasn't excited about this idea. Global warming seemed vague and boring compared to subatomic particles, quantum mechanics, and relativity.

But I thought about it. I did some reading and went to visit some university meteorology departments. It started to sound better. Understanding the atmosphere has a natural, visceral appeal; we all live in it. I liked the emphasis on computer simulation. It seemed exciting to create virtual worlds so complex and realistic that one could learn new things about the real world from them. I had always been fascinated by fluid flows. Though I wasn't yet as concerned about global warming as Marit was, the ozone hole was in the news, and overall, weather and climate did seem as though they were becoming more important to human society. This would be good for future job prospects. And the competition seemed less intense. My Bs in physics meant that I would never get into one of the best graduate programs in that field. I came to understand that the best meteorology programs had fewer applicants than those in physics, and might accept a physics graduate with some Bs on his transcript. I came around to Marit's way of thinking, decided to switch to meteorology, and went to MIT.

Much greater meteorologists have come to the field through career changes bearing at least a passing similarity to mine. Modern atmospheric science really started with one such.

Vilhelm Bjerknes was born in Christiania (now Oslo), Norway, in 1862, to Aletta and Carl Anton Bjerknes. Carl Anton was a mathematical physicist, and Vilhelm moved into the family business. He began his career assisting his father. They worked together on a book that would explain the behavior of electric and magnetic fields by making analogies with fluid dynamics, the flow of liquids and gases. This was a work of interpretation rather than of truly new scientific discovery. Both the electromagnetic and fluid dynamical concepts the Bjerkneses were writing about had been understood for some time. We now call these areas classical physics. This is to make the distinction with the new physics that was beginning to emerge in the late nineteenth and early twentieth centuries. The foundations were being laid at that time for quantum mechanics and relativity. These theories were radical, counterintuitive, and exciting. To some extent, they would prove classical physics wrong.

Electric and magnetic fields had been believed to act through the medium of an invisible, material "ether" that was everywhere. The analogies between fluids and electromagnetism in the Bjerkneses' book were intended in part to prove that ether existed, but the new physics was in the process of proving conclusively that it did not. The career prospects were dimming for someone whose primary achievement would be a book about a hypothetical fluid whose days were numbered.

In 1890, Vilhelm managed to get a job in Bonn, Germany, as an assistant to the experimentalist Heinrich Hertz, whose important research on electricity and magnetism seemed to provide justification for the Bjerkneses' work. (You have heard of Hertz if you have ever looked at the technical jargon in small print on any electrical appliance. Frequency, that is, how often something happens, is measured in a unit that bears his name. One Hertz, or Hz, is one cycle per second. Electrical power is delivered through the wall as alternating current, reversing direction sixty times per second in the United States, or 60 Hz.) Bjerknes did well enough that Hertz hoped to retain him in Germany, but Bjerknes's loyalty to his father brought him home after a couple of years.

Bjerknes's experience with Hertz earned him a faculty position at what is now the University of Stockholm (then called the *Högskola*, literally "high school"), on whose campus I am writing this chapter.

In Stockholm, Bjerknes became friends with colleagues who studied the atmosphere and ocean: Otto Pettersson, Svante Arrhenius, and Nils Ekholm. Arrhenius, who would receive the Nobel Prize in Chemistry in 1903, was the first scientist to propose that human emissions of carbon dioxide would lead to significant warming of the climate. Ekholm worked at the Swedish Weather Service, but also lectured in meteorology at the university. Ekholm saw that while fluid dynamics might not be generating a lot of excitement in physics, someone with Bjerknes's knowledge could make a major contribution to meteorology.

Bjerknes was not terribly interested in weather initially. But time and professional reality wore him down. His work in classical physics wasn't leading him to great recognition, as modern physics rapidly evolved away from him. The meteorologists, on the other hand, continued to show great interest in his work.

In 1898, Bjerknes wrote a paper on what has become known to meteorologists as the Bjerknes circulation theorem. This is a mathematical result derived from analysis of the fundamental equations that govern fluid flow.

Earlier versions of the theorem had been derived by the English physicist Lord Kelvin and the German Hermann von Helmholtz. These theorems said that the circulation (the rate at which a ring of fluid moves around a closed circuit) would remain constant in time in the absence of friction. This meant that vortex motions could neither be created nor destroyed. Bjerknes found that the circulation could in fact change if the density of the fluid were not only a function of pressure (as Kelvin and Helmholtz had assumed) but also of temperature. In a gas, such as the atmosphere, density does depend on temperature: if we compare two balloons filled with equal volumes of air at the same pressure but different temperatures, the colder one will be heavier. Such a fluid is called baroclinic. The extratropical atmosphere, with its strong temperature fronts, is a prime example. The winter storms of Sweden (and the rest of the world outside the tropics) exist because of the baroclinic circulations that Bjerknes's theorem allowed.

The essential mathematical content of the new circulation theorem had been derived two years earlier by the young Polish physicist Ludwik Silberstein. What was new in Bjerknes's work was not the theoretical result itself, but his understanding of the result's relevance to the atmosphere. The meteorologists who read his paper responded enthusiastically.

Bjerknes began to grasp the extent of the open territory in meteorology. Weather prediction was a primitive activity, even by the standards of the day, if compared to physics. Forecasts were made by a combination of experience, intuition, and statistical methods—looking for similarities between the present weather situation and past ones, and using past behavior as a guide to future performance. The results were not good. But the weather had to obey the laws of physics, and those laws were known. Why couldn't one use that knowledge to make forecasts? If so, might the results not be of great value? Maybe studying the atmosphere was a worthy pursuit, Bjerknes reasoned, and a better career option than sticking with the more traditional physicist's path on which he had been.

By 1904, Bjerknes had been thinking about the problem seriously for a few years. He published a paper in the German journal *Meteorologische Zeitschrift*, which laid out the ideas that would form the foundation for modern weather prediction. The paper began:

If, as every scientifically inclined individual believes, atmospheric conditions develop according to natural laws from their precursors, it follows that the necessary and sufficient conditions for a rational solution of the problems of meteorological prediction are the following:

1. The condition of the atmosphere must be known at a specific time with sufficient accuracy
2. The laws must be known, with sufficient accuracy, which determine the development of one weather condition from another.

Bjerknes's first condition required many simultaneous observations, to capture the initial state of the whole atmospheric region for which one was trying to forecast. For Sweden, this meant that one would need simultaneous weather observations over not only the whole country (which, at the time, included what is now Norway) and nearby countries, but also the Atlantic Ocean, since weather systems could move from the ocean to the land during the forecast time. Bjerknes judged that the necessary network of observing stations could be built. Ships could make observations at sea, and the wireless telegraph could allow the data to be collected quickly. Observations of the upper atmosphere as well as the surface would also be needed. Weather balloons had been developed recently,[29] and Bjerknes could see that it should soon be possible to launch these regularly, at least from stations on land.

Bjerknes went on to the second condition: knowing the laws governing the transition from the current weather to the future weather. The relevant laws of physics, expressed as mathematical equations, were known. There were some tricky details that were hard to handle and would have to be ignored for the time being, including the passage of light through the atmosphere and the condensation of water vapor. But he expressed optimism that none of these issues were showstoppers.

The last question was whether one could actually do the necessary calculations to obtain the future atmospheric state (tomorrow's weather) from the observations taken today.

There were seven equations, all but one of them of the type known as partial differential equations. They clearly couldn't be solved exactly, in terms of formulas that one could write down with pencil and paper. The method of solution would have to be approximate. Bjerknes envisioned using graphical methods. One could represent the initial data, obtained from the observations, by carefully drawn diagrams. The rules embodied by the dynamical equations could then be expressed as rules for turning those diagrams into new ones containing the information about the future weather. In 1904 even Bjerknes could not envision the rise of digital computers to do this work.

After Norway became independent from Sweden in 1905, Bjerknes was recruited back home to the University of Kristiania, in Norway, in 1907, then again to the University of Leipzig, in Germany, where in 1912 he became director of the new Leipzig Geophysical Institute. In 1917 the Norwegian government brought him back home again, sponsoring him to set up the Geophysical Institute in Bergen, Norway. The "Bergen School," as it is now often called, became the real birthplace of modern meteorology.

Bjerknes convinced the government to sponsor a set of stations that would make weather observations simultaneously each day. These observations were used to produce daily weather maps, which Bjerknes and his assistants would analyze. They described the high-latitude weather disturbances in their maps as disturbances along the "polar front" separating the cold polar air from the warmer air at lower latitudes. Their work established that the main source of energy for the storms was this temperature contrast. The storms' circulations were driven by the baroclinic effect that Bjerknes's circulation theorem had explained.

Anyone who was anyone in meteorology at the time passed through Bjerknes's lab. The new incarnation of the field grew out of Bergen to become global. The great theorist Carl-Gustaf Rossby began his career there, spending a year and a half with Bjerknes before a stint at the Swedish Weather Service; Rossby would then go to the United States, where he would found the departments of meteorology at both the Massachusetts Institute of Technology and the University of Chicago. (In 1949, Rossby would eventually return to Stockholm. A celebrity in Sweden by that time—he had been on the cover of *Time* magazine—when he founded the Department of Meteorology here, he was able to convince the government to endow it with a generous fund to sponsor visits by foreign scientists. This program still exists today; I am writing this chapter in Stockholm thanks to Rossby.)

Bjerknes's group took meteorology into the twentieth century, but they couldn't yet realize the second part of his vision for physics-based weather prediction. The task of solving the equations accurately and quickly enough to produce a useful forecast would fall to their intellectual descendants.

Battlefield Math

The English mathematician Lewis Fry Richardson would make the first true attempt. Richardson knew of Bjerknes's work and set out to perform the calculations necessary to make a weather forecast using the equations of motion. After several years of preparatory work, Richardson did the math himself, using just a slide rule, over a period of six weeks. He didn't have a nice office; he did it on the French battlefields of World War I, where he was an ambulance driver.

Richardson tried to forecast the pressure change over a six-hour period for one point in central Europe. A six-hour forecast that took six weeks to produce clearly wouldn't be any good to anyone, but the point was to see if it could be done. With digital computers still unforeseeable, Richardson envisioned that if he could do it in six weeks, a future army of human "computers" doing the necessary calculations in parallel could do it quickly enough to produce a useful forecast.

After six weeks of effort, Richardson predicted a pressure increase of 145 hectoPascals in six hours. Standard sea level pressure is 1013 hPa, and typical pressure fluctuations are a few hectoPascals, or maybe 10. A high-pressure system as strong as Richardson predicted has almost certainly never occurred in the history of the earth. The forecast wasn't just wrong, it was impossible.

Strictly speaking, Richardson hadn't made any errors in his calculations; he had set them up wrong. There was more than one way to formulate the equations to pursue an approximate solution, and Richardson had made a poor choice. His method was theoretically correct but unstable. Even a tiny error at the start, of the sort that is unavoidable when using real data, would grow so rapidly as to make the forecast diverge rapidly from reality.

Though Richardson wasn't successful, he had come closer than anyone else before. He published his results in a book, *Weather Prediction by Numerical Process*, which influenced those coming after. But it would take thirty more years for numerical weather prediction to become practical. Several big problems would have to be solved.[30]

First, the observational network would have to be upgraded to the point where enough upper-air measurements in enough places were taken often enough, and the observations collected quickly enough, to provide initial conditions for the calculations in close to real time. Bjerknes worked first with kites; before long, weather balloons would take over.

Second, digital computers would have to be invented. Richardson's idea that human beings could do the calculations quickly enough was completely impractical. He thought that sixty-four-thousand people would suffice to do it. As crazy as that sounds, it was an underestimate; the mathematician and scientific historian Peter Lynch estimates that it would have taken millions.

Third, even if the initial data were there and one had a computer powerful enough to do the calculations, one would need to understand the problem better than Bjerknes or Richardson did. This was not just a matter of mathematical technique, though that would be part of it. Before approximating the equations to make them digestible by the computer, one would have to write down the right equations. The physical laws, in the raw form that they were known, were not manageable. They were right, but they contained too much information.

Signal and Noise: Geophysical Fluid Dynamics

The raw equations governing fluid flow explain not only the evolution of the weather, but also sound waves. Any sound you are hearing as you read this can be expressed mathematically as a solution of the equations of fluid motion for a gas. Sound waves cause the air to vibrate rapidly; the frequencies we can hear are in tens to tens of thousands of cycles per second. A weather system takes days to pass through, so it has a frequency of once per several days. It's very, very slow compared to sound waves. The same equations also contain internal waves. An internal wave is somewhat like a wave on the surface of the ocean, but in the atmosphere. It results from heavier air being lifted up above lighter air; the heavy stuff falls back down. That sets off ripples, just like those on the surface of the lake, except that these can move up and down as well as horizontally; sometimes you can see internal waves manifest in regularly spaced parallel lines of cloud. Internal waves are slow compared to sound waves, but still very fast compared to the large-scale atmospheric disturbances, the high- and low-pressure systems, that cause big day-to-day changes in the weather.

In a weather forecast, one wants to predict just those large, slow-moving weather systems. The sound waves don't matter. A weather forecast doesn't have to tell you what you will hear tomorrow, just whether it will rain. Nor, it turns out, do you need to know about the internal waves. But the equations have all those in them. If one were to solve the raw equations with infinite accuracy, one would in principle reproduce all these high-frequency phenomena as well as the weather.

The accuracy required to do that is unthinkable, even today. The computer needs to solve the equations on a "grid." At each time, each field (temperature, pressure, wind, etc.) is represented by a digitized image, with pixels of a certain size, just like in a digital camera. To resolve all the scales of motion, the pixel size would have to be infinitesimally tiny, making the computation infinitely large. Similarly in time, the equations are stepped forward at an interval, just like the frame rate in a movie. The faster the frame rate, the more expensive the calculation, and the longer it takes to do it. To resolve all the sound waves and internal waves, the frame rate would have to be so high that, even on today's supercomputers (let alone those of decades ago), there would be no hope of doing the calculations quickly enough that the forecast weather would be known before the real weather happened. Useful numerical weather forecasts could be produced only if a way could be found to simplify the equations of motion so that they contained only the low-frequency weather disturbances.

Rossby, in particular, recognized the value of making approximations to the equations in the interest of isolating the most important aspects of the solutions. The most fundamental type of low-frequency oscillation in the extratropical atmosphere is now known as the Rossby wave. It is a sideways "long wave" of alternating cyclonic and anticyclonic rotating disturbances, each often thousands of kilometers long. Rossby waves are set off by any truly large-scale disturbance imposed on the atmosphere. The Rocky Mountains, when the jet stream blows over them, set off Rossby waves: ripples in the jet, thousands of miles across, control the weather downstream, in Europe and beyond. The large trough over the eastern half of the United States and the blocking high over the western Atlantic during the period before and during Sandy's lifetime were parts of a Rossby wave.

The wave was present in the mathematical results of others before, but Rossby made it easier to understand and interpret with a key approximation.

He derived the wave on the atmosphere of a hypothetical flat earth. He retained just one mathematical aspect of the planet's roundness, the one that causes the Coriolis force to divert the horizontal winds more strongly the higher the latitude. That one, the term in his equations representing what is known as the "beta effect," was responsible for the existence of Rossby waves. This peculiar flat earth was impossible, but Rossby knew that this didn't matter. It was the simplest model that would produce the essential result in a way that was correct enough. This combination of an intuitive leap and mathematical elegance set the standard for the new field of geophysical fluid dynamics.

Perhaps the most important triumph of this new field in the twentieth century was the explanation of the basic physics of mid- and high-latitude extratropical storms. These are the baroclinic disturbances that exist due to the instability of the jet stream and the associated pole-to-equator temperature gradient. Bjerknes's group in Bergen had established that the temperature gradient was essential, but they had not been able to formulate a mathematical model that could explain from first principles the storms' existence and properties. Without such a model, the Bergen School's work remained less than fully predictive.

The problem was solved approximately simultaneously, in two different but complementary ways, by an Englishman, Eric Eady, and an American, Jule Charney. Both Eady and Charney came to understand the baroclinic weather systems as resulting from the interaction of the temperature gradient at the surface with disturbances at upper levels. Their two theoretical models were different in some details but shared most of their essential features. Both made elegant approximations, very much in the vein of Rossby, to simplify the mathematics. Both ignored the fact that the earth is a sphere and worked in a local Cartesian geometry, as though the surface of the earth were flat, as Rossby had. Just as important, both sets of equations filtered out internal waves. Both models predicted correctly the observed scale and structure of the highs and lows familiar to all extratropical meteorologists.

In both Eady's and Charney's models, a storm could grow stronger only if the low-pressure trough tilted westward with height, against the shear, so that a low-pressure system at the level of the jet stream started out to the west of a low-pressure system at the surface. Each low-pressure center is encircled by cyclonic winds (counterclockwise in the Northern Hemisphere), in turn embedded in the larger-scale sheared "steering" flow associated with the jet stream. We still recognize this westward tilt as essential. As the shear means the eastward steering winds aloft are faster than those at the surface, the upper low is blown eastward faster than the surface low, and brought closer to it, reducing the tilt and bringing the system closer to vertical alignment. As this happens, the atmosphere below the upper low is lifted, pulling in air at the surface toward the center and accelerating the surface cyclone's winds as the high angular momentum is brought inward. Though Eady and Charney didn't envision a weather system in which the surface low-pressure system was a tropical cyclone, the basic mechanism works the same way in that case, and all this was happening to Sandy in 2012.

Eady's paper was published in 1949. Charney published his in 1947, the year after he finished his PhD in mathematics at the University of California, Los Angeles. After that, like meteorologists of past decades, Charney planned to spend time in Norway. He got a fellowship to spend a year in Oslo (where Bjerknes would pass the last few years of his career until his death in 1951) to work with Bjerknes's associate, the theorist Halvor Solberg. On his way there, he stopped in Chicago to visit Rossby. The two got along so well that Rossby convinced Charney to put off his Norwegian fellowship and stay to work with him in Chicago for a year.

Around this time, the first digital computers were being built. The physicist and mathematician John von Neumann had become aware of the first general-purpose computer, the Electronic Numerical Integrator and Computer (ENIAC). He had commandeered it for his work on the design of the hydrogen bomb. After the war, von Neumann was at the Institute for Advanced Study in Princeton, New Jersey, where a new computer was being built for him. Von Neumann realized that weather prediction would be a good problem for a computer to solve, and communicated with Rossby about it. Rossby set up a meeting between von Neumann and about twenty meteorologists, and made sure Charney was there. Impressed by Charney, von Neumann recruited him, upon his return from Norway, to lead what would be the first truly successful effort in numerical weather prediction.

Charney and several other meteorologists would spend three years at Princeton. Von Neumann's computer didn't exist yet, but there were theoretical problems to solve first. They worked on those.

Charney built on his theoretical model of baroclinic weather disturbances to generate a set of equations that could be used to generate weather forecasts. He derived that set, the so-called quasi-geostrophic equations, by a mathematical method known as asymptotic expansion. Successful use of this method requires not just mathematical expertise but physical intuition. One has to know a lot about what the solutions to the equations will look like before having derived the equations themselves. Charney could make the asymptotic expansion work because he understood not just mathematics but also the physics of the weather. He had absorbed the insights of the Bergen School, of Rossby, and others.

The quasi-geostrophic equations are accurate enough to capture a wide range of weather situations with a useful degree of accuracy, as long as one doesn't try to use them too close to the equator. Equally important, they contain no internal waves or any other high-frequency solutions; their only waves are the slow Rossby waves. This made them much more practical to solve on the computer.

Because the high frequencies were filtered out of the equations, the time step in the numerical solution procedure could be much longer than if internal waves were resolved. This made it possible, for the first time, to do the computation faster than the weather itself evolved.

Charney had actually derived the quasi-geostrophic model before coming to Princeton, and published it in 1948. But additional problems had to be solved before it would be possible to set up the numerical forecasts. One was that three spatial dimensions represented still too big a problem for the computer that would be available to him. The weather forecast problem had to be reduced to a two-dimensional one, in which the whole atmosphere would be represented by the flow at a single horizontal level: a single weather map at one altitude, rather than a stack of them representing the flow different heights aboveground. Another analysis of the equations of motion indicated that the best level to represent all the others would be one in the middle, around 500 hPa, about three miles up.

With all the preparatory work done, von Neumann's computer was still not ready. Unwilling to wait, von Neumann managed to get time for the weather prediction project on the existing ENIAC machine. The results were successful. A forecast good enough to be useful was produced, using the laws of physics, in a time short enough that future forecasts could be made before the weather actually happened. After less than half a century, Bjerknes's vision was reality.

After Charney and colleagues published their results in 1950, their methods were quickly accepted by the larger meteorological community. The Princeton efforts led directly to the adoption of numerical prediction by the U.S. Weather Service. The forecasts we have today are direct descendants of the first ones done at Princeton.

The details have all changed. It is no longer necessary to use the quasi-geostrophic model, for one thing. Today's computers are enormously more powerful than ENIAC, have no problem resolving internal waves and other high-frequency motions, and can use more accurate equations. This allows the forecasts to be global, including the tropics, where the quasi-geostrophic model would fail. The observational network is much better than it was in 1949. The advent of satellites, and improvements in weather balloons, surface observations, and other technologies, allow better knowledge of the atmosphere's current state, a necessary prerequisite for a good forecast. Everything about the models, about the way the observations are used to derive the initial conditions, and about every other step in the procedure is tremendously more sophisticated than it was sixty-five years ago, after decades of research into both the physics of the atmosphere and the mathematics of solving the equations. But the essential idea comes from Bjerknes, as implemented by Charney and his colleagues at Princeton: prediction based on the physical laws that govern the atmosphere.

Chaos Theory

The most important conceptual advance in the years since the Princeton project was one that would define the limits (at the time, nowhere near in sight) of what the weather prediction enterprise could ever hope to achieve. In 1961, Charney was chair of the Department of Meteorology at MIT, the department started by Rossby decades earlier. His colleague Ed Lorenz was doing numerical calculations of his own.

Lorenz's model was much simpler even than the single-level quasi-geostrophic model of the Princeton group of over a decade before. Lorenz had started from the equations representing the flow that occurs when a layer of water sits between two parallel plates and the bottom one is heated while the top one is cooled, generating a circulation as the warm water rises and the cool sinks. Already an exquisitely crude model for real weather, it was boiled down yet further. Lorenz used severe mathematical approximations to reduce this system down to equations to predict just twelve numbers, each varying in time. Twelve is a very, very small number compared to the number of variables needed for real weather prediction.

Like the original fluid equations from which he started, Lorenz's equations were nonlinear. This means that multiplying the inputs by two, say, wouldn't necessarily lead to outputs also two times larger. Nonlinear equations are inherently much more difficult than linear ones, and their properties much more poorly understood. Even simple nonlinear equations are difficult to solve by hand, so Lorenz needed a computer. By 1961, computers had already improved a lot since ENIAC, but they were still extremely primitive by today's standards. Lorenz's severe approximations were necessary to allow him to solve his equations on a computer modest enough so that he could have it to himself.

Lorenz became puzzled. He had repeated the same calculation twice, expecting the new one to match the old one precisely for its entire duration. Instead, while the two calculations agreed at the start, the new one drifted apart from the old one, eventually giving completely different results. He noticed eventually that the starting data had not been input to the new calculation with exactly the same precision—the same numbers of digits after the decimal points—as in the old one. The difference was in the fourth decimal place—no more than a few parts in ten thousand.

This difference grew rapidly, so that the two solutions soon became as different as if their starting points had been entirely different, instead of almost identical. Lorenz came to realize that this was not an error, but instead a real property of his equations. They exhibited what we now know as deterministic chaos (or just chaos, for short).

Small differences in initial conditions grow exponentially rapidly in a chaotic system. Even the tiniest errors in one's knowledge of the initial state amplify so quickly that, before long, the state of the system will be as unrelated to the initial conditions as if it were behaving randomly. Since the initial conditions are all one knows before making a forecast, such a system *cannot be forecast* beyond the point where this effective randomness sets in.

This is so even though the dynamics of the system, as represented by the equations, is deterministic, meaning the same inputs always lead to the same outputs. The problem is that one can never guarantee that two real inputs are truly exactly the same. The demonstration that a deterministic system could behave as though it were random meant that some phenomena could be truly unpredictable even with effectively perfect knowledge of their dynamics. This was as important a discovery as quantum mechanics or relativity.

The phenomenon of deterministic chaos had been discovered by the French mathematician Henri Poincaré in the late nineteenth century, but not much had come of it. What Lorenz came to realize was that not only were his simple mathematical systems chaotic, but that the real atmosphere almost certainly was as well. This meant that there was a fundamental limit to how far in the future one could ever predict the weather. Lorenz estimated that this limit was around two weeks, a number still believed to be approximately correct.

In 1963, when Lorenz published his first paper on his results, entitled "Deterministic Nonperiodic Flow," in the *Journal of the Atmospheric Sciences*, what he had discovered was not an issue yet for real weather predictions. The weather forecasts of the time (for which Lorenz's simpler calculations were idealized proxies) were nowhere near good enough to approach the predictability barrier. Even at one or two days, they were often substantially wrong; a two-week forecast was unthinkable even without worrying about chaos.

But the progress in weather forecast skills since then has been steady. Today, ten-day forecasts can have some skill, and deterministic chaos is a big factor in weather forecasting. Much of the research effort in numerical weather prediction in the last couple of decades has focused on dealing with it.

Modern weather forecasters deal with chaos mainly by using "ensemble" methods. The idea was implicit in Lorenz's first glimmer of understanding of the problem.

Lorenz found that two forecasts will eventually become very different if their initial conditions are even slightly different. But the initial conditions are never known precisely. The initial conditions are based on measurements of the atmosphere's current state. These have errors, and even very small errors will grow in a chaotic system. Besides the errors in the measurements themselves, those measurements are at only specific places; in between those places, we have no information. Whatever we assume about those in-between areas may be wrong; this is known as a sampling error.

Both measurement and sampling errors mean that multiple initial conditions are possible, as far as we know. As the differences grow rapidly, we can't tell which of the multiple forecasts is right.

Ensemble methods deal with this problem simply by making all those different forecasts, or at least as many of them as possible. The different ensemble members, like Lorenz's original calculations, use the same model but are generated from slightly different initial conditions, each consistent with the actual observations of the atmosphere, within the known uncertainties. At the start of the forecast, the different ensemble members will be close to one another, because the initial conditions were. As the model runs progress, representing times farther in the future, the ensemble members will start to drift apart. How quickly they drift apart tells the forecaster something about the degree of uncertainty in the forecast.

Each ensemble member defines a weather situation that could occur, as far as the forecaster knows. If all the members stay close together for a few days, it's pretty certain that the weather will be similar to all of them during that time. If they spread far apart, as in the case of a hurricane going either right or left, the uncertainty is much greater.

Much of the science in ensemble forecasting lies in figuring out how to generate the ensemble members in the most efficient and intelligent way so that a modest number of forecast model runs can accurately represent the full range of possibilities and indicate (for example, by majority vote) which is the most likely. Figuring out how to do this efficiently is a complex problem in applied mathematics, atmospheric physics, and software engineering.

The result of all the work that has been done to solve these problems is that we have good weather forecasts—better today than ever. Forecast skill has steadily improved, decade after decade, continuing to the present. The improvement is due not to any single breakthrough (post-Lorenz), but to great effort, by many people, in many areas: the science of both observational and theoretical meteorology, the mathematics of numerical computation, and increasing computer power, which itself is a result of steady advances in many areas of science and engineering. Most of this work, it is worth pointing out in our current political climate, was government-funded.

Some people still like to complain about forecasts being unreliable. They are simply mistaken in this. Perhaps they remember only the occasional busts and forget all the forecasts that were right. Or perhaps they have unrealistic expectations, such as a prediction days ahead of time of the precise timing of a rain shower. When it comes to the large-scale weather, our forecasts are getting to the point where the Lorenz limit comes within view; seven- or even ten-day forecasts now have some skill. There is still room to improve further, but not nearly as much as there used to be.

Graph showing the steady improvement in U.S. weather forecasts over time. The *x* axis is a particular numerical measure of the skill in the forecasts of geopotential height at 500 hPa over North America; the *y* axis is the year in which the forecasts were made (from 1955 to 2005). The upper curve shows thirty-six-hour forecasts, while the lower curve shows seventy-two-hour forecasts. Along any given horizontal line, the crossings of the two curves are separated by around twenty years, indicating that the seventy-two-hour forecast at any given time has been roughly as good as the thirty-six-hour forecast twenty years ago was. *(Courtesy of the National Centers for Environmental Prediction/NOAA)*[31]

Without the prior century of advances in weather forecasting, the forecasts of Sandy in 2012 would have been inconceivable. It would have been impossible to predict, days ahead of time, that the storm would turn left and strike the coast while moving westward. No forecaster had ever seen that occur before, since no storm had ever done that, and no statistical model trained on past behavior would have produced it as a likely outcome. In fact, forecasters not only could see that outcome as a possibility over a week ahead, but were also quite confident of it by four or five days ahead.

Many other phenomena we might like to predict are probably not susceptible to the approach that has worked for weather. Baseball games, elections, and economies, for example, behave as they do because of the collective behavior of human beings. It seems unlikely that we will ever truly know the laws that govern human behavior as well as we know the physical laws that govern the atmosphere. Even with other physical systems, prediction can be stymied by difficulties that meteorologists haven't had to face. Earthquakes, for example, depend on the behavior of the solid earth, which is much harder both to observe (we can't send weather balloons down into the earth's crust) and to understand (the laws governing geologic faulting are not known as well as those governing normal fluid flow).

But even if the achievements in modern meteorology can't be repeated in other fields, we should still recognize them for what they are. Forecasts like the ones we had as Sandy formed and moved up the coast don't come from the heavens. They're the result of a century of remarkable scientific achievement.

The Tropics

Tropical cyclones were not part of this story. The geophysical fluid dynamics that led to numerical weather prediction was relevant only outside the tropics. Tropical weather was terra incognita to Bjerknes, and would remain poorly understood for decades beyond the first numerical forecasts in Princeton.

Hurricane forecasting would develop as a specialized field, to some degree separate from the rest of weather forecasting and atmospheric science. The influence of the military in providing new technologies was important, particularly in the early years.

Aircraft reconnaissance of hurricanes began in 1943, when Col. Joseph Duckworth and Lt. Ralph O'Hair, apparently on a whim, flew into one from their wartime flying school in Bryan, Texas.[32] Aircraft observations have played a critical role in both scientific research and forecasting of hurricanes from then up to the present, and are still made routinely by the air force (as well as NOAA). Radar, another military invention, would allow the first direct visualization of the hurricane inner core structure (eye, eyewall, rainbands) decades before satellites became available.

Until recently, standard numerical weather prediction models were not able to capture tropical cyclones. Their resolution was too coarse to simulate the eyewalls, rainbands, and other narrow features. The resolution was far too coarse to simulate cumulus clouds, critical to hurricane dynamics. The collective effects of the clouds had to be represented in an implicit way, by pieces of computer code called parameterizations. These parameterizations were inadequate, introducing additional errors into the model predictions.

Unable to rely on the numerical weather models, hurricane specialists developed their own models as aids to forecasting. Some were statistical, making predictions of a current storm based on what similar ones had done in the past. Others were dynamical (that is, based on the equations of motion as with a normal numerical weather prediction model), but different from the standard numerical weather models of the time in that they were designed only to predict hurricane tracks, rather than to produce comprehensive weather predictions. One involved dropping a mathematical idealization of a prefabricated hurricane, one too small and intense for the weather models of the day to have produced on their own, into the observed large-scale environmental winds, allowing researchers to see which way the winds would take the storm.[33]

The 1990s brought a watershed for hurricane prediction. The resolution and physics of true dynamical models, the descendants of the Princeton quasi-geostrophic prototype, had improved to the point where such models started to show some skill at predicting the tracks of hurricanes. They could do this, however, only once the hurricanes were already there in the initial conditions. They couldn't yet predict the formation of a hurricane (genesis) in the first place.

Sometime in the 2000s, the models began to do a plausible job of genesis as well. Just the year after Sandy, 2013, NHC started making predictions of tropical cyclogenesis five days in advance, rather than just two days, as in the past. In large part, this reflects the center's new confidence in the models.

Tropical weather still works differently from extratropical weather, and always will. Tropical cyclones in particular still present unique challenges. Forecasts of tropical cyclone intensity, in particular, have not improved much in decades, though there are signs that this may be starting to change.

New technologies, such as direct incorporation of Doppler radar data from aircraft into specialized, high-resolution dynamical models, are showing promise under sustained effort from many scientists.

The earliest pioneers of physics-based weather prediction did not address these problems at all. Living in high latitudes, they did not think much about tropical cyclones. But the consequences of their vision have come, finally, to include the tropics fully. Hurricane forecasting still requires special expertise, but it is becoming more and more similar to "regular" weather forecasting. The models have become the central tool for both.

The relevance of the early work in geophysical fluid dynamics to Sandy, however, was still a bit more direct. Charney and Eady had formulated the quasi-geostrophic equations, necessary for the first forecast, to explain the essential mechanism by which baroclinic weather systems grow. Their equations described deep lifting and strengthening of a surface cyclone as an upper trough is moved over it by shear. This process was happening to Sandy on Friday. The storm was becoming less of a hurricane and more of a hybrid. The numerical models were capturing the transition with crystal clarity.

12

FRIDAY, OCTOBER 26

The Two Faces of Shear

By the morning of Friday, October 26, Sandy had passed the central Bahamas. The northern islands of Great Abaco and Grand Bahama were under a hurricane warning; tropical storm warnings were in place for the rest of the northern Bahamas and for much of the coasts of both northern Florida and the Carolinas.

In satellite imagery, Sandy's appearance had become more asymmetric, a result of the shear and of the interaction with the upper low during the previous day. A spiral core of cloud was still apparent around the center, but it was thin, and dwarfed by an enormously broad arc of thick cloud to the north and northwest of the center, filling the entire enormous area of ocean between the storm and the continent, and extending onshore into Florida, Georgia, and the Carolinas.

Forecaster Richard Pasch's 11:00 a.m. discussion on Friday first commented on displacement of the storm's center of circulation from most of the convection, attributing that to both the shear and the dry air wrapping around from the south. Pasch estimated a current maximum surface wind speed of seventy knots (still in hurricane territory), and then reported on the latest forecast model simulations:

DYNAMICAL MODELS SHOW AN ADDITIONAL INCREASE IN SHEAR OVER THE NEXT 24 HOURS . . . WHICH SHOULD CAUSE AT LEAST

SLIGHT WEAKENING. LATER IN THE FORECAST PERIOD . . . GLOBAL
MODELS SHOW SOME RE-INTENSIFICATION OF THE CYCLONE . . .
WHICH IS VERY LIKELY DUE TO BAROCLINIC INFLUENCES.

Visible satellite image of Sandy from 5:00 p.m. EDT on October 26. This is a "super rapid
scan" image from the NOAA GOES-14 satellite, provided by NASA.

The word *baroclinic* describes, literally, a situation in which surfaces of
constant pressure are not parallel to surfaces of constant density. The two
kinds of surfaces cross each other at angles.[34] More simply, a baroclinic atmosphere is one in which temperature varies in the horizontal. The phrase
"baroclinic influences" means influences associated with horizontal temperature gradients, such as those between the tropics and the pole.

On a spinning planet, temperature gradients are related to winds in a
particular way.

Atmospheric pressure is itself just the weight of the atmosphere above. Warm air is less dense than cold air at the same pressure. Pressure gradients and winds are close to geostrophic—winds blow along the contours on the pressure map faster where the contours are more tightly packed—especially outside the tropics. Because of these three facts, horizontal temperature gradients and vertical wind shear go together. If there is one, there must be the other. This is known as the thermal wind relation. Every student of meteorology learns it early and is never allowed to forget it.

The shear is such that the wind increases with height in a direction *perpendicular* to the temperature gradient. The jet stream itself exemplifies this. In most places, the atmosphere gets colder as one goes from equator to pole. The thermal wind relation implies that the wind should become more westerly (eastward) with height. The jet stream does this. It is a belt of strong westerly winds at high altitude, with weaker winds below it: westerly shear. The African easterly jet, on which the easterly wave that became Sandy formed, also obeys thermal wind. That jet has strong easterly winds aloft, implying easterly shear; that is consistent with the reversed temperature gradient in West Africa, with the hot Sahara lying north of the cooler coast.

The jet stream, and the horizontal temperature gradients that go with it, are unstable, as explained by Bjerknes, Charney, and Eady. The disturbances that result from the instability are extratropical weather systems. These are known in the business, in fact, as baroclinic disturbances. The shear and the temperature gradients constitute one package.

When Pasch wrote the phrase "baroclinic processes," he was using shorthand for all that. He was indicating that the same processes that strengthen normal winter storms would also strengthen Sandy in a couple of days.

Just in his previous sentence, though, he indicated that before that happened, vertical shear would weaken the storm. This made sense; shear normally weakens hurricanes. But "vertical shear" and "baroclinic processes" are almost synonyms for each other. Pasch seemed to be saying that the same thing would weaken the storm and then strengthen it. In fact, he *was* saying that. But he used different words to telegraph the different dynamical roles played by shear in different types of systems, Sandy possessing some of the qualities of both.

The hurricane Sandy, still retaining some of its warm core and its tropical convection, would be weakened by the shear, as it would get dry air in closer to the storm. But eventually the "baroclinic processes" (the face of the shear that strengthens extratropical cyclones) would kick in and strengthen the storm. This would change its character:

BY 72 HOURS . . . THE MODELS SHOW SIGNIFICANT COLD AIR ADVECTION OVER THE SOUTHWEST PORTION OF THE CIR-CULATION . . . WHICH SHOULD HASTEN THE EXTRATROPICAL TRANSITION PROCESS.

The word *advection* means "blowing in from nearby." So the models predicted that upper-level cold air from the north (an indicator of an upper-level low-pressure trough) would blow over Sandy's southwest quadrant. This would drive an upper-level pressure drop and strengthen Sandy's cyclonic circulation. This is a typical process by which an extratropical cyclone strengthens, just as in Eady's and Charney's models. So it would "hasten the extratropical transition process."

Pasch went on to describe uncertainty in the timing of the extratropical transition:

THE OFFICIAL FORECAST SHOWS THE SYSTEM AS POST-TROPICAL AT 96 HOURS . . . THOUGH THERE IS SOME UNCERTAINTY AS TO THE TIMING OF THE TRANSITION. REGARDLESS . . . WHETHER SANDY IS OFFICIALLY TROPICAL OR POST-TROPICAL AT LANDFALL WILL HAVE LITTLE BEARING ON THE IMPACTS.

In other words, it wasn't exactly clear when Sandy would officially make the transition to "post-tropical," meaning, a formerly tropical cyclone that had become fully extratropical. NHC had to make a forecast as to when that would happen, regardless; its responsibility was tropical cyclones, and if the storm were to become post-tropical, it would leave NHC's jurisdiction. Pasch was indicating that it was a tough call. It was even tougher than he indicated, to some extent, because it always is; there really isn't a hard line between tropical and extratropical storms.

Usually, it doesn't matter. Most extratropical transitions occur out at sea or when a storm is weakening or is already weak. Though neither was forecast to be the case for Sandy, Pasch indicated that, really, it still shouldn't matter exactly where the line was drawn between tropical and extratropical. Such a distinction would "have little bearing on the impacts."

Hurricane versus Not

It's critically important to understand this. It confused many important people before landfall, and many in the media continued to get it wrong well afterward. I am determined that you, reader, get it right.

The distinction between a tropical cyclone and an extratropical cyclone is not that one is more powerful and another less so. It is a structural difference, a change in type. The features distinguishing one type from the other are technical. I will explain them, but if you don't understand or don't remember them, that's okay. What you have to remember is that, in some circumstances, an extratropical cyclone can be just as powerful as a tropical cyclone. Thus, *if a tropical cyclone changes into an extratropical cyclone, that does not necessarily mean it has become weaker or less dangerous.*

It is true that the most powerful hurricanes have stronger winds than any extratropical cyclone; no winter storm can match a category four or five hurricane. But a strong extratropical cyclone can certainly have winds equivalent to a category one hurricane, as Sandy was on Friday, October 26. And it is likely to be much larger.

As I said, the distinction between a tropical and extratropical cyclone is a structural one. It's a difference in what type of storm it is, not how strong it is. A tropical cyclone may weaken as it transitions to post-tropical, but it may not. In the case of Sandy, it was forecast not to weaken, and the forecasts proved right.

Think of the difference between male and female athletes. The fastest male sprinter in the world can beat the fastest female one, but even the slowest woman on any country's Olympic track team can smoke most men alive in the 100-meter. The analogy, admittedly, would be better if a runner could change his gender during a race. But the point is that a change of type from one storm to another wouldn't mean that Sandy would have to weaken as it transitioned, because a strong post-tropical cyclone can be just as powerful as a hurricane. The storm would change at some point from warm core to cold core, and from retaining some circular symmetry to being completely asymmetric. But its winds wouldn't necessarily slow down. In fact, because the radius of high winds could expand even farther, the storm might become still more destructive as it transitioned.

The complexity of the situation, and the degree to which Sandy was becoming a difficult forecast communication challenge—even as what the storm would actually do was being quite confidently and consistently predicted by the models and the forecasters—became more evident as the hours passed. On the night of Friday the twenty-sixth, forecaster Jack Beven wrote the discussion. His first paragraph began:

SANDY IS SHOWING CHARACTERISTICS OF A HYBRID CYCLONE THIS EVENING. OVERALL . . . THE SYSTEM LOOKS LIKE A LARGE OCCLUDED FRONTAL LOW. HOWEVER . . . SURFACE OBSERVATIONS DO NOT SHOW STRONG TEMPERATURE GRADIENTS . . . AND CENTRAL CONVECTION IS OCCURRING IN BANDS TO THE NORTHWEST OF THE CENTER. IN ADDITION . . . AMSU DATA FROM NEAR 2000 UTC INDICATED THAT THE SYSTEM STILL HAD A DEEP WARM CORE.

An "occluded frontal low" is an extratropical, baroclinic system in which, typically late in life, the surface cold front and warm front have collided. A weaker temperature contrast usually remains along the occluded front at the surface. So the absence of strong temperature gradients in the surface observations was a little inconsistent with the characteristics of an occluded frontal low that Beven noted in his first sentence.

Beven mentioned also a measurement from the AMSU, the Advanced Microwave Sounding Unit, an instrument on a polar-orbiting satellite. Polar orbiters see only narrow swaths of the atmosphere at a time, and observe a hurricane at any given time only by luck. By using microwave radiation, though, they can see through clouds to take the temperature underneath. The AMSU told Beven that Sandy still had a warm core, a characteristic of a tropical cyclone. Like the lack of any apparent surface front, this seemed to contradict the otherwise hybrid appearance of Sandy, presumably gathered from geostationary satellite images showing a strongly asymmetric cloud pattern.

Beven went on to elaborate on the forecast for the next couple of days leading up to the expected landfall. It hadn't changed much in twelve hours, but the details continued to come into focus. The storm was under strong shear. This would keep weakening it, as one would expect with a hurricane. Interaction with the upper trough to the southwest (the typical process that intensifies extratropical storms) was keeping the system going despite the shear, but it was expected to dip below hurricane strength soon. Sandy would turn northeastward, and continue that way for a while.

But then the new shortwave trough—the one farther north, riding along the giant jet stream dip associated with the strong negative NAO—entered the picture. This second extratropical system would simultaneously whip Sandy westward, and strengthen it, again through the same extratropical, baroclinic processes. The storm was forecast to regain hurricane intensity. At the same time, its identity as a hurricane (a tropical cyclone, with the structural characteristics that this definition entails) was forecast to become steadily more doubtful. Beven's discussion closed by repeating that while the meteorological situation was hard to categorize, it didn't really matter:

THE INTENSITY FORECAST SHOWS SANDY REGAINING HURRI-CANE STRENGTH IN 48–72 HR. HOWEVER . . . THE CYCLONE WILL BE UNDERGOING EXTRATROPICAL TRANSITION AS THIS HAP-PENS . . . AND WHEN THIS PROCESS WILL BE COMPLETE IS UNCER-TAIN. REGARDLESS OF THE EXACT STRUCTURE AT LANDFALL . . . SANDY IS EXPECTED TO BE A LARGE AND POWERFUL CYCLONE WITH SIGNIFICANT IMPACTS EXTENDING WELL AWAY FROM THE LOCATION OF THE CENTER.

The process by which the shortwave trough was forecast to pull Sandy onshore before merging with it has a name, though Beven didn't use it: the Fujiwhara effect. When two typhoons in the Pacific, for example, come close enough, they can influence each other's tracks. They orbit around each other, for at least part of a rotation. Here the same thing was happening in the models, just with two storms of different types. I had never seen this before, certainly not in real time, and certainly not over my own city. It was thrilling; but the "significant impacts extending away from the location of the center" were terrifying.

13

FUJIWHARA

A vortex is a flow pattern in a fluid that has rotation about a center: water spiraling down the bathtub drain, the swirling eddies made by a canoe paddle, a hurricane.

A marker dropped into the flow near a vortex (a cork dropped into water, for example) will orbit about the vortex center while spiraling in toward it. At the same time, a vortex itself can also behave like a marker dropped into the fluid and can move, like a cork, with the larger-scale flow in which it's embedded. If you paddle your canoe in a river, the vortices your paddle leaves behind will float downstream just like a cork would.

If two vortices come close enough to each other to get caught in each other's flows, then each one acts like the cork; its center moves in the flow swirling around the other one. At the same time, if one is moving, the center about which the other is moving is itself moving, and vice versa, and the two carry out a joint maneuver.

How the dance goes depends on the two vortices' directions of rotation and their relative strengths. If they are spinning opposite ways but are identical in every other way, they will move in parallel straight lines, perpendicular to the line through the two vortex centers. If they are exactly identical, spinning the same way at the same rate, they will circle each other, both orbiting about the point between them. (If the strengths are different, they will orbit about the "center of mass," a point that will still lie on the line between them, but closer to the stronger one.) This do-si-do is known as the Fujiwhara effect.

Sakuhei Fujiwhara was a Japanese meteorologist working in the early twentieth century. After a few years in the Japanese weather service, he got his PhD in 1915, and went on to study with Vilhelm Bjerknes in Bergen, Norway. At the time, this was the meteorological capital of the world. Bjerknes and his students had begun the process of turning weather forecasting into a modern science; spending time in Bergen was the only way to get to the frontier of the field. Fujiwhara also spent time at Imperial College in London, before going back to Japan in 1920.

Fujiwhara did experiments in water tanks in which he explored the interaction of vortices; it appears he did these both in Bergen and after getting back to Japan. He published the results in a paper in 1923, in the *Quarterly Journal of the Royal Meteorological Society*.[35] The first part of the paper contains descriptions of the results of many experiments: Fujiwhara made bigger vortices and smaller vortices, some rotating the same way, some opposite ways, and described all the different trajectories the vortices would take. In many cases, the vortices would merge with each other in the end. Two strong vortices rotating in the same direction would produce a stronger one after the merger.

After some theoretical analysis, Fujiwhara went on to describe the relevance of the results to real weather systems. These included a merger of two tornadoes over South Wales, a merger of two extratropical storms over the United States in 1921, and other similar cases over Japan.

The interaction of Sandy with the extratropical shortwave trough was forecast to play out exactly as Fujiwhara's original paper described it.

Sandy and the extratropical system both had strong winds in midtroposphere, around 500 hectoPascals (three miles or so aboveground). The effect was very clear in sequences of maps of the forecast model winds at this level. As the trough moved east and came within Sandy's flow, the mutual rotation brought the trough down to the south while bringing Sandy north. As both Sandy's flow and the larger-scale eastward jet stream pulled the trough eastward, it aligned to Sandy's south. Sandy was then pinned between the trough to the south, with its own counterclockwise flow, and the blocking high centered to the northeast, with its clockwise flow. In the models, the combination would catapult Sandy westward into the coast, as the trough spiraled around and into Sandy, finally being absorbed just during landfall.

Occasionally, two tropical cyclones get close enough to each other to do the Fujiwhara dance. Most often, this happens in the western Pacific, where there are more tropical cyclones: around thirty per year. It happens only rarely in the Atlantic, where the annual total is closer to ten, making a close encounter of two much less likely.

In this case, it was being forecast to happen with a tropical cyclone and an extratropical cyclone. This Fujiwhara interaction would lead to Sandy's final extratropical transition into a "post-tropical" storm; to a westward turn, never observed before in this region; and to a catastrophic landfall.

The models visualized this days ahead of time, almost exactly as it would happen. I saw it, along with every other meteorologist who was paying close attention, in the model output.

Many atmospheric scientists are drawn to the field by its direct relevance to human experience. Still, for those of us whose primary role is as research-ers or teachers, our relationship to the subject matter is, most of the time, impersonal. Genuine excitement can grow out of intellectual fascination, but is disconnected from the firsthand circumstances of our own lives. Most of the time, an academic researcher in my field is no different, in this way, from one who studies galaxies or zebrafish. Those of us who study severe weather events do, on the other hand, become fascinated by atmospheric disturbances that are also human disasters. The popularity of fictional weather disaster movies, not to mention The Weather Channel, indicates that we are not alone in this.

For scientists like me, the death, misery, and destruction wrought by a hurricane or tornado are, of course, part of why we do what we do. If we can understand these storms better, maybe we'll be able to predict and pre-pare for them a little better. Maybe that will help, next time, somewhere. We tell that to ourselves, and our colleagues, and funding agencies. It's true enough. A grisly death toll or economic damage total is as good a justifi-cation for research as anything. But studying a major weather disaster can still feel, sometimes, a little like rubbernecking at a nasty traffic accident. One is drawn in, compelled to understand the specifics, fascinated—but still detached.

Both the fascination and the detachment are necessary ingredients for the practice of science. No matter how serious the patient's condition, no matter how gruesome the symptoms or bleak the outlook for survival, the doctor has to be both dispassionate and intellectually engaged to provide the best treatment. The analogy isn't perfect, because meteorologists can't ever "cure" anything in the way that a doctor can sometimes cure a patient. The best we can do is make a prediction—and researchers like me don't even do that; we only try to understand. Still, understanding a disaster is the same as understanding anything else. One first has to want to understand it, which requires an intellectual passion. This in itself carries a kind of emotion. But one also has to be able to look at the data as objectively, as dispassionately, as possible.

When the site of the disaster is where you live, it's different. But it also isn't. One's home and one's community are threatened. That induces concern, anxiety, fear. But the nature of the threat, in all its detail, is still fascinating. If one is worth anything as a scientist, one is not going to be able to turn that off.

The Fujiwhara effect is a meteorological curiosity. It's not the subject of important research. We have understood it since the 1920s. There are no conferences on it; it is not featured in special issues of meteorological journals. It happens from time to time, usually far away, and that's it—but it's beautiful to watch.

It has poetry to it. Two giant entities in the atmosphere, dangerous and powerful but elemental, not easily anthropomorphized (even when we give them human names), and most of all normally solitary, each doing its own thing, engage with each other. Like interplanetary bodies coming within each other's gravitational pull, or dancers engaging after finding each other across a crowded room, or fighters trying to take each other's measure before a big bout, they circle; then they close in, and then become one. This phenomenon has an exotic name that no one but us weather geeks knows, but the phenomenon itself appeals to everyone. When I talk in a grade school or at a science fair, it always comes up. Some kid always asks if two hurricanes can crash into each other. And what is a scientist but a big kid, really?

In the model predictions, the Fujiwhara interaction of Sandy, the mutual rotation and merger with the shortwave trough, would give the storm a new jolt of energy just as it would make landfall, with the center south of New York City. Since a cyclone's winds are counterclockwise around the center in the Northern Hemisphere, a landfall point to the south meant that the winds would be blowing toward the northern Jersey Shore and into New York Harbor. An onshore wind is what causes the worst storm surge.

The southern landfall also put the city on the storm's right flank, where the forward motion of the storm itself would be in the same direction as the cyclonic wind on the right, the two adding up to make a stronger total wind felt on the ground.

The storm was also forecast by the models to be extremely large at landfall. The wind field had expanded in the earlier trough interaction near the Bahamas. The next interaction being forecast by the models, the final extratropical transition as a result of the collision with the shortwave trough, would expand it further. A larger area of high winds also means a higher storm surge, as the larger "fetch" (the distance over which the wind blows in the same direction over the water) gives the long wind-driven wave more time to build.

So the combination of the landfall point to the south and the large size of the storm meant that if the model predictions bore out, the situation would be just right for a massive storm surge. I'm not a storm surge expert, but I knew this much.

I also knew what the potential consequences were for New York City. They had been well documented, actually, in the scientific literature, in government reports, and in the popular media. Katrina, seven years earlier, had given the United States a graphic example of what hurricane flooding can do. New York City is neither as topographically challenged nor as hurricane-prone as New Orleans, but there were enough parallels. New York City also has many low-lying areas, is surrounded by water, and sits in a location that, scientists and government officials long knew, could be reached by a hurricane. The right hurricane, combined with the shape of the harbor, could produce an overwhelming storm surge. And the population, economy, and infrastructure are much greater than those of New Orleans. The potential for complete destruction might not have been as great in New York City as it was in New Orleans, but there was more there to destroy.

I knew some of the details. I knew about the vulnerability of the subway system in particular, because my colleagues Klaus Jacob, George Deodatis, and their students and associates had recently written a detailed report on it for New York State.[36] I hadn't read the report, but I had heard Jacob describe the basic conclusions: The subways would flood in a major storm surge. They weren't protected from that in any way, and they weren't designed to recover quickly. Every New Yorker understands clearly that without the subways, the city would be paralyzed and dysfunctional. I also understood that large areas would be likely to lose power; certainly this would happen outside the city, and in some of the less densely populated areas of the outer boroughs, where aboveground power lines are standard and easily knocked over by falling trees. But even inside the city it could happen, if critical electrical infrastructure got soaked. There was concern about saturation of power lines underground in Manhattan. I understood that with the subways out and power down over large areas, many other challenges would cascade down from these.

I wasn't scared for my own life or health, or those of my family. We live on a hill far above the Hudson River, so there was little chance that our apartment building would flood. The winds at landfall were forecast to be in the category one range—enough that you definitely shouldn't be outside, but the city is mostly made of heavy brick, concrete, and steel buildings that aren't going to be severely damaged by category one winds.

We might lose power for a few days. If that happened we would have water for a couple of days, from the tank on the roof, but the pump that refills that is electric, so we would run out of water if power were out longer than that. I knew we had better stock up on water, food, and batteries; shortages in shops were certainly possible. We hoped our gas stove would still work. We could lose our electronic communications. The cell phone network could fail, and so could the Internet, and that would be unpleasant for my family, addicts all. (We gave up our land line years ago.)

These things, individually, would be just inconveniences. (In fact, none of them would happen to us, in our neighborhood, but they all happened to huge numbers of people.) Even if the worst happened and we were to experience all these things, we would still almost certainly survive, unharmed, after a period of stress and hassle.

But New York is a city of eight million people. Many of them were much more vulnerable than we were. Those in the lowest-lying areas stood a good chance of having their homes and businesses flooded by storm surge. It was reasonably likely that some people, possibly many, would be injured or killed if the water got that high. And if the subways were to flood and power to go out over large areas, the city would be truly crippled. I didn't try to think through all the consequences, but just thinking that far was enough to scare me.

Our city, even more so than most, is a tightly connected, carefully or-chestrated machine with many moving parts. A lot of things have to go right to make a normal day in New York City possible. The attacks of 9/11 had demonstrated the machine's fragility. Sandy probably wouldn't kill as many people as bin Laden had, but it could do much more widespread harm to our complex infrastructure. I knew all this. As the storm drew closer and I saw the Fujiwhara effect play out in the forecast models, I was excited to see all the aspects of this event, looking more and more historic, play out, in the atmosphere and the ocean. But as the forecast became more certain, and it became clearer that a real disaster was not just a remote possibility but a likelihood, I began to feel fear as well.

It was not just me. Heading into the weekend, the storm drew closer, and the forecasts became more certain. The reality of the threat to New York, New Jersey, and the surrounding states began to crystallize. The machinery of emergency management began to shift into gear. Government officials at the local, state, and federal levels had decisions to make. They had to make them in the face of a storm without precedent. The American system of storm warning and response had not been built for Sandy.

14

SATURDAY, OCTOBER 27

On Saturday, October 27, at 5:00 p.m. EDT, Sandy's center was fixed by NHC at 30.2 degrees north, 75.2 degrees west, well offshore of the Carolinas and moving toward the Northeast.

The maximum sustained winds were estimated at sixty-five knots (seventy-five miles per hour), keeping Sandy just barely a category one hurricane. The central pressure minimum, however, at 961 millibars, was abnormally deep for a storm of this intensity. This reflected Sandy's size.

The wind depends on the pressure gradient, or the rate at which pressure drops over distance. A traverse from Sandy's center to the point at which its winds would drop below tropical storm force would span 520 miles. This long distance could accommodate the large pressure drop with winds weaker than would be found in a smaller storm of the same central pressure. The total energy in the circulation, though, was enormous. Sandy's tropical storm–force winds swirled over a patch of sea huge enough to hold several normal hurricanes.

Tropical storm warnings were in effect for Bermuda and for the Carolina coast between South Santee River, South Carolina, and Duck, North Carolina, in the Outer Banks near the Virginia border. A tropical storm watch, the lower-level alert, held farther south in South Carolina, between the Savannah and South Santee Rivers, where any threat from Sandy would soon pass.

The tropical storm warnings stopped at Duck. NHC issued no formal watch or warning for Virginia or anywhere north—but not because those regions were safe.

The National Weather Service had decided that no tropical watches or warnings would be issued north of Duck because Sandy was forecast to be "post-tropical" by the time it made landfall. This decision had been considered for several days, and was finally made in a conference call the previous day among the directors of the Hydrometeorological Prediction Center (which has primary responsibility for predicting extreme rainfall), the Ocean Prediction Center, NHC, and the heads of the local weather service offices in Morehead City, North Carolina, and Sterling, Virginia.[37]

The decision was made because Sandy was expected to become an extratropical storm, outside NHC's jurisdiction. This meant that the primary responsibility for forecasting the storm and its consequences would lie with the local National Weather Service offices. The intent of the decision was to control the time and place at which this handoff of responsibility would occur. In this way the NWS would avoid confusing emergency managers, government officials, and the public with midstream changes in the types of forecasts, watches, and warnings being issued.

By Friday, in fact, before the time of the final decision, the local office in Mount Holly, New Jersey, had already issued a gale watch on Sandy. A gale watch is a nontropical product. It indicates increased likelihood of gale-force winds, meaning those over thirty-four knots. Thirty-four knots is also the threshold for declaring a tropical storm. But a declaration of the existence of a tropical storm requires that other criteria be satisfied as well: a warm core, some degree of circular symmetry (or at least lack of temperature fronts), convection near the center. NHC makes that call. If it had made it—that is, if it expected Sandy to have tropical characteristics when reaching New Jersey—a tropical storm watch rather than a gale watch would have been in effect. Although the implication of the potential wind strength was the same, Mount Holly's gale watch implied a nontropical system.

But Sandy was still a hurricane in the present, and NHC's advisories and discussions continued accordingly. The 5:00 p.m. advisory, written by forecaster Daniel Brown, after describing the current state of the storm and the forecast track, included assessments of the hazards to those on land. A hurricane can pose threats due to wind, storm surge, and rainfall. These were addressed one at a time:

HAZARDS AFFECTING LAND

WIND . . . TROPICAL STORM CONDITIONS ARE EXPECTED TO REACH THE TROPICAL STORM WARNING AREA IN THE CAROLINAS WITHIN THE NEXT FEW HOURS . . . AND SPREAD NORTHWARD TONIGHT AND SUNDAY. TROPICAL STORM CONDITIONS ARE POSSIBLE IN THE WATCH AREAS THROUGH TONIGHT. GALE FORCE WINDS ARE EXPECTED TO ARRIVE ALONG PORTIONS OF THE MID-ATLANTIC COAST BY LATE SUNDAY OR SUNDAY NIGHT AND REACH LONG ISLAND AND SOUTHERN NEW ENGLAND BY MONDAY MORNING. WINDS TO NEAR HURRICANE FORCE COULD REACH THE MID-ATLANTIC STATES . . . INCLUDING LONG ISLAND . . . BY LATE MONDAY.

The careful wording of this statement reflected the transition from tropical to post-tropical that was expected north of Duck. "Tropical storm conditions" were expected in those areas with tropical storm warnings and watches, while further north "gale force winds"—exactly the same as tropical storm–force winds, but resulting from a nominally different type of storm—were expected farther north. The mid-Atlantic states could even see "hurricane force" winds. They would not see an actual hurricane, according to the best scientific judgment available; not that it would matter.

The risks of storm surge were predicted quantitatively in those places where significant surge was expected within the next forty-eight hours. Forecasts of storm surge are inherently more uncertain than those of the storm itself, and thus are predicted only two, rather than five, days in advance. For the first time, on the preceding day, at 11:00 p.m. EDT, New York, New Jersey, and southern New England had come within the forty-eight-hour horizon.

STORM SURGE . . . THE COMBINATION OF A DANGEROUS STORM SURGE AND THE TIDE WILL CAUSE NORMALLY DRY AREAS NEAR THE COAST TO BE FLOODED BY RISING WATERS. THE WATER COULD REACH THE FOLLOWING DEPTHS ABOVE GROUND IF THE PEAK SURGE OCCURS AT THE TIME OF HIGH TIDE . . .

FL EAST COAST NORTH OF CAPE CANAVERAL . . . 1 TO 2 FT

NC SOUTH OF SURF CITY . . . 1 TO 3 FT

NC NORTH OF SURF CITY INCLUDING PAMLICO/ALBERMARLE SNDS . . . 3 TO 5 FT

SE VA AND DELMARVA INCLUDING LOWER CHESAPEAKE BAY . . . 2 TO 4 FT

UPPER AND MIDDLE CHESAPEAKE BAY . . . 1 TO 2 FT

OCEAN CITY MD TO THE CT/RI BORDER . . . 4 TO 8 FT

LONG ISLAND SOUND . . . RARITAN BAY . . . AND DELAWARE BAY . . . 4 TO 8 FT

The forecast didn't explicitly say, "New York Harbor," but the mention of Long Island Sound and Raritan Bay implicitly included it.

The forecast of four to eight feet was exactly the same as had been given, one day ahead of time, for Hurricane Irene the previous year. When Irene actually made landfall in Brooklyn, the surge was at the low end of the forecast range. The storm surge, or elevation of the water due only to the storm, was a little over four feet. The total water elevation at the Battery in Lower Manhattan, relative to low tide, was nine and a half feet. This was enough to put water just over the tops of the lowest seawalls. It was not enough to cause serious flooding, though it came extremely close. Having made it over seawalls, each additional foot of surge would have been another foot of water "above ground."

Sandy's storm surge would measure nine feet at the Battery. The height of the water above the mean tide would be eleven and a half feet; the height above low tide, nearly fourteen feet.

What Is Storm Surge?

Storm surge is the persistent elevation of the sea level caused by winds. It is not the tides, which normally raise and lower sea level twice per day. It is the rise of the sea—temporary, weather-induced, but often slower than the tide—above wherever the tide would otherwise have it. The total water level, surge plus tide, is called the storm tide. Levels can be measured relative to the typical low tide (mean level of low water, or MLLW) or to the average midpoint between low and high tide. When Sandy made landfall at the Battery, the tide was high, about five feet above MLLW. The surge was nine feet, so the storm tide was nine plus five, or fourteen feet above MLLW.

Then there are waves. Waves pile water on top of its average level, higher than the storm tide. Because each wave comes and goes quickly, waves do not cause persistent flooding in the way that storm surge does. But they can deliver water with great force, and cause great damage to structures exposed to them.

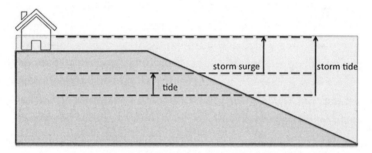

Diagram illustrating the distinction between the (astronomical) tide, storm surge, and storm tide. The lowest dashed line represents the reference level, which could be either the average water level or the mean level of low water (MLLW)—the typical low tide. The astronomical tide is what the sea level would be at a given instant in the absence of any weather disturbance. The storm surge is the difference between the astronomical tide and the actual sea level, that difference being due to meteorological disturbance. The storm tide is the total water level above the reference, astronomical tide plus surge.

The numbers given by NHC are stated as water depths "above ground." They are actually calculations of storm surge. But it's too hard to predict the timing exactly enough to pinpoint whether the tide will be high or low as the surge peaks. It's conservative to assume that the peak surge will occur at high tide. If the lowest-lying spots on the coast are close to flooding at a normal high tide, then a surge of five feet will put those spots five feet underwater.

To a typical layperson, though, and perhaps even to an emergency manager trained to handle severe weather, what "above ground" means is not clear. "Above ground" where? Where I live or somewhere else? The advisory itself doesn't specify; it can't, really, without becoming encyclopedically long, since one would need to give a number for every spot on the map. Because of the multiple ways storm surge can be reported (storm surge or storm tide, relative to low tide or to mean tide?), and because most people don't know the elevation of their homes or businesses very accurately, the numbers in the forecasts are not easily interpreted by those not used to reading them or hearing them reported in the media. In New York, New Jersey, and New England, major storm surge events are rare, so almost no one there, experts aside, knows what they mean.

The forecasters at NHC had been aware of this difficulty for some time. Their storm surge team was already working on a new product to fix it. They were designing a high-resolution inundation map (a color-coded map showing how deep the water could actually be, down to street level), to be issued for every place threatened by surge. But this product was still in development, not ready to be issued yet. The old numbers, stated in text advisories, would have to do.

What Causes Storm Surge and How Is It Forecast?

When the wind blows over the water, it exerts a force. The same thing happens when a solid surface moves relative to the air; this is the air resistance that slows down your car, or an airplane, or a tennis serve. Here, it is the wind that is slowed down, while the water is pushed, generating a current in the direction of the wind. As the wind becomes stronger, it also churns up larger waves. These are corrugations to the surface of the water, and obstacles to the wind. The wind exerts extra pressure on the windward sides of the waves, thus pushing them downwind—and in the process, exerting yet more force on the water. Stronger wind churns up bigger waves, which more effectively absorb the wind's momentum. More wind stress[38] leads to more waves, which leads to more wind stress, dragging the wind down farther but accelerating the upper ocean's motion.

Earth's gravity pulls the ocean toward the planet's center, holding it in place like tea in a cup. The only major force pushing it in horizontal directions (sideways) is the gravity of the moon and the sun. This force is not steady. As the moon orbits the earth, the earth spins, and the pair of them orbit the sun, the global ocean sloshes back and forth. This creates the tides. The tides are experienced at the coast as periodic (once or twice daily) raisings and lowerings of sea level, as gravitational forces push water onshore and then off again. The detailed phases of the tides can vary in complex ways from place to place, as the water has to flow over and around the intricate contours of the sea bottom and coastlines. But once known, the tides are as predictable as the celestial orbits that produce them.

When the winds of a strong storm pull on the ocean's surface, that stress can move the water, just as the tides do. But instead of reversing twice a day, the winds push steadily in the same direction for as long as they blow that way. As the water moves in response to the wind, it will become deeper downwind, where it accumulates, and shallower upwind, from where it is coming. If the winds are blowing toward shore, water will move onto the land. That is storm surge. In some cases, the surge can come in very quickly. Although the causes are different, a storm surge is quite similar to a tsunami in its basic physics—both are elevations of the water surface that travel inland essentially as very long waves. The experience of the two phenomena on the ground can be similarly terrifying.[39]

While wind is usually the main cause of storm surge, surface air pressure also plays a role. The water rises slightly in areas of low pressure and is depressed in areas of high pressure. Near the center of a strong cyclone, the surface air pressure is low, and that increases the surge. This is normally a small effect compared to the wind, but not always so small. Sandy's pressure was quite low, and this could have added as much as two feet to the surge.[40]

Predicting the actual height that a given storm surge will reach can be complex and difficult, even though the essence of the phenomenon is simple. How high the water gets will depend on the strength and direction of the wind. It also depends on the upwind-downwind distance, or "fetch," over which the wind blows steadily in the same direction; the same wind over a greater fetch will yield a larger surge, as the water has a longer distance over which to pile up. All these aspects of the wind field (strength, direction, fetch) have to be forecast accurately to produce an accurate surge forecast.

Like the tides, the surge also depends on the topography of the sea bottom. Deep water right up to the coast leads to a smaller surge, while a large shelf of shallow water leads to a bigger one. The surge also depends on the shape of the coastline. New York Harbor sits at the point of a triangular piece of oceanic real estate; water moving landward will be funneled toward the point by the New Jersey coast and the South Shore of Long Island, causing greater accumulation in the harbor.

Storm surge is predicted by computer models that are separate from those that predict the weather that drives the surge. A storm surge model is a virtual slab of water with wind pushing on the top of it, in a small region right near some coastal area that is potentially at risk. The model knows the details of the sea bottom and coastline in great detail, and has very high resolution (small pixels) near the coast to capture them. The model may or may not explicitly include tides; the SLOSH model (for Sea, Lake, and Overland Surges from Hurricanes), used by NHC, does not. Many surge models, including SLOSH, are two-dimensional; they assume that the ocean moves as a true slab, with the same currents at all depths down to the bottom.

Ignoring tides and treating the ocean as two-dimensional may seem like crude approximations. But for forecasting purposes, these are all minor issues compared with the uncertainties in the wind. Because the winds in a hurricane blow roughly in a circle, they change direction 180 degrees as one traverses the center. So if a storm's position is a little different from that forecast, places near the center will experience winds not just a little bit different from what was forecast, but in the opposite direction altogether. This can spell the difference between a big surge and none at all. More generally, the surge is very sensitive to the details of the wind: how strong it is, in what direction is it heading, over how great a fetch and over how long a time, and these in turn are very sensitive to the forecast track, intensity, and size of the storm.

The surge problem, once the wind is known, is relatively linear. Changes in output (surge) are roughly proportional to changes in input (wind). There is no chaos in surge itself; two surge forecasts with slightly different initial conditions, but the same winds, will not give very different results. But there is chaos in the prediction of the winds themselves, and that is multiplied by the sensitive dependence of surge on wind. Given the uncertainties in typical hurricane forecasts, surge forecasts are normally felt to be too unreliable to be issued more than a couple of days ahead of time.

But Not Knowing Doesn't Mean Not Being Concerned

While an accurate forecast of storm surge may not be possible more than a couple of days ahead of time, it is definitely possible, at least in some cases, to see much earlier if the *potential* for a large surge exists. Sandy was such a case.

It had been clear as early as a week before Sandy's landfall, after the first numerical model runs showed scenarios similar to what eventually happened, that there was at least the *potential* for a catastrophic event somewhere on the northern East Coast of the United States. At that early time, it wasn't possible to be certain that this potential would be realized; as late as Wednesday, October 24, a significant fraction of the models was taking the storm out to sea. By a day later, though, on Thursday the twenty-fifth, the great majority of the models was predicting landfall, and NHC's forecast had also locked in on that conclusion. The location of the landfall was still uncertain, but New York and New Jersey were definitely well within the range of possible targets. Over the succeeding days, that range continued to narrow steadily. The storm surge forecasts from Friday evening showed that the New York metropolitan area was directly in danger.

Even at the early times, though—really, from the very first vision of Sandy's landfall in the models on October 22—it was clear not just that there could be a landfall of a major storm, but that the impacts could be enormous if the forecast came true. Major coastal flooding induced by storm surge in particular was a clear threat. The *probability* of the threat was very uncertain. It was not unreasonable to view the probability as low at those early times. One couldn't be at all certain that anything bad would actually happen. At the same time, the *nature* of the threat—what could happen, if something did—was clear.

In his blog post of Tuesday, October 22, Jason Samenow of the *Washington Post* had written (italics in original)[41]:

> Based on current information, I'd give slightly better than 50 percent odds that a significant storm impacts *some place* in the mid-Atlantic and/or Northeast early next week,

Under the heading "What's the worst case scenario?" he went on to write:

For the mid-Atlantic, it would probably resemble current model simulation of the Canadian model or last night's European model. In these scenarios, what's now tropical depression 18 intensifies into a hurricane—named Sandy—near the Bahamas and then combines with a powerful mid-lattiude [*sic*] cold front (upper level trough) approaching the East Coast. The result is a super-intense, slow-moving "hybrid storm"—as powerful as a major hurricane. Potential effects include:

- tremendous coastal flooding (exacerbated by a full moon and higher than normal tides)
- copious rainfall and damaging winds extending a few hundred miles inland—including the Washington, Baltimore, Philadelphia and New York City regions

For the most part, this is exactly what happened, spelled out in clear language, just about a week ahead of time. Samenow gave fifty-fifty odds of it happening somewhere; of course this meant smaller odds than that for any particular location. He went on to give a 20 percent chance of it happening in the Washington, DC, area, with farther north more likely than farther south.

Samenow got it about as dead right as anyone could have at that point. Others might have given greater or lesser odds, or explained it less clearly, but he wasn't the only one outlining essentially the same scenario. There is no question that information was broadly available a week ahead about what *might* happen. The quantity and certainty of that information had increased steadily throughout the week.

On Wednesday, October 24, I was interviewed by a local television station, Fox 5. The reporter and I sat in my office and looked at my computer screen. We examined the "spaghetti plot," the different model tracks shown together on a map. One could see from this map that both a landfall and a recurving path out to sea were still possibilities. One of the tracks was the deterministic forecast from the ECMWF model. The deterministic forecast is made with the European Centre's best model, run at the highest resolution, costing the most computer power. (The ensemble runs that produce the spaghetti plots are made from a cheaper, lower-resolution version. These cheaper models can be run many times to generate the multiple tracks necessary to account for initial condition uncertainty amplified by chaos.) That model, even more than in its solutions from the previous couple of days, showed a scenario much like what eventually happened.

The reporter and I also looked at the contour plot of surface pressure for the five-day forecast. It showed a very deep low-pressure center making land-fall just south of New York City. Winds blow counterclockwise around a low in the Northern Hemisphere, so the location south of New York City meant that strong winds would be blowing directly onshore. This was not good.

Predicting storm surge accurately is a complicated scientific problem, but the basic phenomenon is relatively simple. The wind pushes the water hard enough that a lot of it moves in whatever direction the wind is blowing. If it's blowing toward the shore, then water moves toward the shore. When the water gets close to shore, where the bottom is shallow, it piles up and spills onto the land. Strong onshore winds mean a high storm surge. That's about it.

I'm being slightly facetious. I've just told you that even given precise knowledge of the winds, how high the water gets depends on many subtle factors. These include the bathymetry (the depth and shape of the ocean bottom) near the coast, the topography of the land just inland of the coast, and of course whether the tide is high or low when the surge arrives.

At five days' lead time, one has no hope of trying to evaluate all that, because the winds aren't known accurately enough yet. That's why the numerical storm surge models that are used to produce quantitative predictions of the surge aren't run until two or three days before landfall. One can't know the exact timing of landfall five days ahead, either, and that will determine whether the surge hits at high tide or low. In the case of Sandy, the tide tables said that the tidal range would be five feet, so the difference between landfall at high and low tide would be enormous. We (I, the reporter in my office, and everyone else looking at the same information or any other information that existed at that time) had no way to know which would occur, or even if there would be a landfall near us at all.

But we could see plausible scenarios. The ECMWF model, with its reputation as the best one in the world at that moment, was tracking a major storm into the coast just south of us. That storm was blowing the winds onshore with hurricane force, right into New York Harbor. I'm not a storm surge specialist, but I knew enough to know "hurricane force wind onshore = bad."

I told the reporter this—that there was a significant risk of a high storm surge if the ECMWF deterministic model were to turn out to be correct. But because of the disagreement among the models, with some tracks in the spaghetti plot curving out to sea, I told him that the event also "could be nothing"—no storm at all for New York, or for anywhere in the U.S.

I think I said about the right thing in that interview, but that didn't reflect any particular genius on my part. It was a straightforward statement of the overall consensus view. The reporter could have gotten a similar statement from any knowledgeable scientist looking at publicly available information. Indeed, many other sources in the media were saying approximately the same thing, and had been for two days. (NHC, at the time I was speaking to the Fox 5 reporter, wasn't yet forecasting landfall, just because it was still too early. It doesn't forecast further than one hundred twenty hours ahead. Although landfall wasn't much more than one hundred twenty hours from that interview on the twenty-fifth, NHC forecast that Sandy would progress along its track slightly slower than it eventually did. By the next morning, Thursday the twenty-sixth, NHC was predicting landfall on Tuesday morning, slightly later in time but very close in space to where it actually occurred.)

By Thursday the twenty-fifth, the wide spread in the spaghetti plot was beginning to collapse. The models that had been taking Sandy out to sea, such as the American GFS model, were starting to blink, and bringing it back into shore with the others. The majority of the models, and NHC, were now predicting landfall. There was still a great deal of disagreement in where the landfall was predicted to occur, and about the timing and intensity of the storm at landfall. So there was still no reason to panic in New York—but there was a clear *possibility* of a very severe event. And the probability was increasing, as indicated by the graph showing the probabilities of high wind speeds near New York City forecast by several weather models over time, shown in the color insert.

On the morning of that same day, Joe Lhota, head of the Metropolitan Transportation Authority, was interviewed on WNYC, a New York City affiliate station of National Public Radio, by Brian Lehrer, a seasoned and respected radio and television journalist. They covered many topics; I assume that the interview had been scheduled well before, and not specifically because of Sandy. But Lehrer brought up the forecast, and asked Lhota if he thought they would need to close the subways, as they had during Irene the year before. Lhota said, speaking about Irene (which had happened before his time as head of the agency):

"The subway system had never been shut down before . . . and it was a good exercise. I don't think we're looking at anything like that for what's happening next week, but we are going to have all indications are, a pretty brutal nor'easter. What's gonna happen north of New York City is I think where the most problems are gonna be, because they're talking about the fact that while it may be rain and wind down here, not that far north of us, they're going to be having snow, so it's going to be a replica of last year's holiday snowstorm."

Lhota was not worried about flooding, and didn't think there would be any need to shut down the subways. With little uncertainty ("all indications are"), he anticipated a milder event than Irene had been. I don't expect the MTA head to be a meteorologist, but I assume he had been talking to some. Whatever message was communicated to him and however that communication was structured, something must have been missing. The fact that, at that point, there *was* certainly a clear risk of major flooding from storm surge must have gotten lost before it got to the head of the MTA.

The next day, Friday morning, I was interviewed by Lehrer on the same radio show, with Sandy now the specific subject. The forecast was becoming still more certain. After asking me a number of questions about the meteorology of the storm, Lehrer played me the recording of Lhota's statement from the day before, and asked me what I thought of it.

I said that while there was still uncertainty in both the track and intensity, successive forecasts had brought Sandy into the coast farther south, putting New York City on the right side of the storm where the winds were likely to be strongest. I said, "It would seem to me that the risk of a big storm surge is pretty significant, and so I wouldn't be surprised if the city re-evaluates over the next couple days whether they might want to take some action based on that."

What I said there, as two days before, was not a result of any deeply original thought. I believe it was more or less the mainstream view of the meteorological community. The NHC forecast at that time was not bringing the storm farther north; it was bringing it quite close to the city. And while there wasn't yet an explicit storm surge forecast, any knowledgeable meteorologist—and there are many such out there, and many of them were talking to the media—could look at the maps and see that there might be a severe storm surge. Why hadn't that message gotten to Lhota?

There is a more direct clue in the city's own formal communications with the public. The record suggests that the confusion about the nature of the storm may have delayed the preparations for the storm in New York. On Saturday evening, Mayor Bloomberg gave a press conference. He explained why no evacuations had been ordered yet:

> Let me tell you first we are not ordering any evacuations as of this time for any parts of the city. We're making that decision based on the nature of the storm. Although we're expecting a large surge of water, it is not expected to be a tropical storm or hurricane-type surge. With this storm, we'll likely see a slow pileup of water rather than a sudden surge, which is what you would expect with a hurricane, and which we saw with Irene 14 months ago.

The mayor said that they were expecting "a large surge of water." Nonetheless, he didn't see that as adequate to justify evacuations. Why not?

By this time, there were quantitative forecasts of the storm surge. The numbers NHC was giving on Saturday (four to eight feet) were low compared to what eventually happened. Perhaps part of the explanation lay in that. Bloomberg would have been justified, based on the best information at that time, in thinking that the surge wouldn't be as large as it turned out to be.

I don't think that's the explanation. In the first place, four to eight feet is still enough to be very serious. Irene's surge had been just over four, and while that had not been a major disaster for the city, it had been very, very close—six or seven or eight feet would have been a major disaster indeed, and Bloomberg had to have known that. In the second place, blaming the low surge forecast just isn't consistent with Bloomberg's words. He said the surge would not be "tropical storm or hurricane-type." He then repeated that the "slow pileup of water" would not be "what you would expect with a hurricane," and implied, as Lhota had, that the surge would not be as bad as Irene's had been.

One can guess, at least in hindsight, what must have been going on here. The description of Sandy as something that would be other than a "tropical storm or hurricane" at landfall was, arguably, meteorologically correct, given the forecast. And whether correct or not, it had been discussed in the media quite a lot already by that point. Sandy was going to merge with an extratropical, "winter storm," to acquire some of the characteristics of both. It wouldn't be a hurricane; it would be a hybrid, a "superstorm" or "megastorm."

Superstorm and *megastorm* are big, scary-sounding names. Somehow, though, the lack of the tropical label that a hurricane warning would have provided was confusing people, even those in the Bloomberg administration, one of the most competent, scientifically serious, and best-prepared local government administrations in the United States, and possibly the world, with respect both to disaster management and to weather- and climate-related hazards in particular.

By 5:00 p.m. Eastern Daylight Time on Saturday, the NHC advisory followed the surge forecast (still four to eight feet for the New York City area) with a statement designed specifically to dispel the kind of confusion Bloomberg's statement had shown:

SURGE-RELATED FLOODING DEPENDS ON THE RELATIVE TIMING OF THE SURGE AND THE TIDAL CYCLE . . . AND CAN VARY GREATLY OVER SHORT DISTANCES. GIVEN THE LARGE WIND FIELD ASSOCIATED WITH SANDY . . . ELEVATED WATER LEVELS COULD SPAN MULTIPLE TIDE CYCLES RESULTING IN REPEATED AND EXTENDED PERIODS OF COASTAL AND BAYSIDE FLOOD-ING. IN ADDITION . . . ELEVATED WATERS COULD OCCUR FAR RE-MOVED FROM THE CENTER OF SANDY. FURTHERMORE . . . THESE CONDITIONS WILL OCCUR REGARDLESS OF WHETHER SANDY IS A TROPICAL OR POST-TROPICAL CYCLONE.

This language is clear. But it appeared in the advisory, a relatively long statement, several paragraphs from the top. Despite it being in all caps, really, it was fine print. It was like a can of food whose label, a flashy one- or two-word name in big print, says one thing while the list of ingredients, in tiny letters read only by careful shoppers, says another. The fine print said that the storm had the potential to be catastrophic. But the label on the can didn't read, "Hurricane," and this diminished the overall impression of the threat.

There was no reason to think the forecast hybrid character would make the flooding any less destructive. The model predictions were now, with much greater certainty, of strong onshore winds at landfall. While the exact landfall location was uncertain, Sandy was forecast to be exceptionally large, with strong winds covering an enormous area. This meant that it was not important to pinpoint the exact location. A large area was at risk. The huge size of the storm also meant a worse surge than would be expected for a smaller storm with the same wind speed. The wind would be acting over an enormous fetch, a factor well understood to increase storm surge.

15

COMMUNICATING THE THREAT:
THE LABEL ON THE CAN VERSUS THE FINE PRINT

I've explained the differences and similarities between a tropical cyclone and an extratropical storm. But it's worth summarizing them again, and then trying to understand how the U.S. system of producing forecasts and warnings deals with those similarities and differences.

According to the NHC webpage, "A tropical cyclone is a rotating, organized system of clouds and thunderstorms that originates over tropical or subtropical waters and has a closed low-level circulation." Other properties that are applied in practice to distinguish tropical cyclones from extratropical storms (a.k.a. winter storms) are that, first, tropical cyclones are warm in the center compared to their environments, whereas extratropical ones are cold; and second, tropical cyclones, at least when mature and intense, have a high degree of circular symmetry, whereas extratropical storms are much more asymmetric. The difference in structure is apparent in satellite cloud images, with mature hurricanes exhibiting the iconic pinwheel shape, with the eye in the middle, while extratropical storms' clouds typically have a "comma" shape.

When an Atlantic tropical cyclone's maximum sustained winds (the strongest winds anywhere in the storm that last at least a minute) reach sixty-five knots (seventy-five miles per hour, or thirty-three meters per second), it has reached category one on the Saffir-Simpson scale, and is called a hurricane. If it keeps its tropical character but the winds weaken below that level, it is downgraded to a tropical storm. This typically happens at or before landfall at the latitudes of the mid-Atlantic or higher-latitude portions of the U.S. East Coast. The storm is moving over cooler waters as it moves north, depleting its energy source, and once it moves over land, it is completely cut off from that source. Thus, normally, when a hurricane goes from being a hurricane to being not a hurricane, that is an indication of weakening. It has become a tropical storm, which is just a weaker form of a hurricane. It is a demotion.

Some northward-moving tropical cyclones begin to interact with extratropical weather disturbances (disturbances in the jet stream, fronts, and other systems typical of the higher latitudes). Sometimes, as with Sandy, the tropical cyclone can encounter a separate, distinct extratropical storm and merge entirely with it. When this happens, the tropical cyclone changes its character to acquire some or all of the traits of an extratropical weather system. It can become asymmetric and cold-cored instead of symmetric and warm-cored. This process is called extratropical transition. It refers to a change in the *structure* of the storm, and implies *nothing about its intensity*. Sometimes storms undergoing extratropical transition weaken during and after the transition; sometimes they strengthen.

It's true that the strongest hurricanes' winds reach intensities greater than that any extratropical cyclone can achieve. But there is a considerable middle range, in which the strongest extratropical cyclones' intensities overlap those of category one or even category two hurricanes. And because extratropical storms (or hybrids, or "post-tropical" storms) tend to be larger than hurricanes, and because storm surge depends not just on the strength of the wind but also on the fetch over which it blows, a tropical cyclone that has become post-tropical can produce a disproportionately large surge, compared to the typical hurricane of the same maximum wind speed. Sandy was the example of this to end all examples.

Sandy's estimated winds at landfall would be sixty-five knots—just barely equivalent to those of a category one hurricane, according to the Saffir-Simpson Hurricane Wind Scale, adopted by NHC in 2009. That scale was a simplification of the earlier Saffir-Simpson Hurricane Intensity Scale used earlier. The earlier scale measured not just wind, but also central pressure and surge. The problem with that earlier scale was that the same wind can be associated with different pressures and surges, making the scale confusing; hence the switch to the pure wind scale. However, had the previous scale been in use, *Sandy would have had a surge of category three, and a central pressure approaching category four, at landfall.*

To make the issue still more complex, extratropical transition is a continuous and often gradual process. There can be an extended period when the storm is a hybrid, with some properties of an extratropical system and some properties of a tropical one, and it is not entirely clear what to call it. But NHC has to make a judgment call at some point, because its job changes once the system is, or is forecast to become, post-tropical.

Through 2012, the rules under which NHC operated said that it could issue a hurricane warning only for a storm that was forecast to remain a hurricane (a tropical cyclone of category one intensity) at the time for which the warning was to apply. If a storm were forecast to undergo extratropical transition, to the point that it was to become no longer tropical, the issuing of a hurricane warning for that time would no longer be warranted, *even if the storm were forecast to be just as destructive as a major hurricane.* Once that happens, the storm becomes the responsibility of the local National Weather Service offices. They can issue their own statements, advisories, and warnings, but not a hurricane warning.

If NHC had issued hurricane warnings, its rules would have required it to stop issuing them once the storm was determined to have become post-tropical, even if it maintained its strength or intensified. NHC, and emergency managers with whom it communicates, believed that this would be more confusing and disruptive than not issuing the warnings in the first place. So they chose not to issue them for any location north of the Virginia–North Carolina border.

A hurricane warning from NHC carries a great deal of weight. Many decisions are triggered when there is a hurricane warning (including some insurance policies). It's simple and clear, and gets attention in a way that nothing else does. It's an alarm, and it either sounds or it doesn't. In Sandy, it didn't. From the point of view of helping people (including government officials) understand what was coming, it should have, because no one other than meteorologists and weather geeks cared whether Sandy was tropical or post-tropical at the time of landfall. They cared what was going to happen to them and their cities, towns, homes, and businesses.

NHC did continue to issue forecasts and advisories on Sandy. These were, for the most part, highly accurate. They made statements, such as the one I quoted earlier, that explained quite clearly what was going to happen. They said that Sandy would be post-tropical, but they also said that this didn't matter; the flooding could be just as bad regardless. They said this early on, and they continued to say it. Eventually, Bloomberg and everyone else understood it. But up through the evening of Saturday the twenty-seventh, the record strongly suggests that the tropical/post-tropical issue was confusing him and his team in the city. If Sandy would be as bad as a hurricane, why wasn't NHC saying it was a hurricane? The rules NHC was operating under said that it couldn't put the word *hurricane* on the label, but that it could say in the fine print on the back of the can that the storm would be essentially the same thing as a hurricane, for all practical purposes. One can see how that might not be easy to understand.

In a presentation at the Annual Meeting of the American Meteorological Society in January 2013, forecaster Bryan Norcross of The Weather Channel said that when he heard Bloomberg's statement in his Saturday evening press conference, about how the surge would not be so bad because it would not be a "hurricane-type" surge, "In my time, and I've been doing hurricanes a long, long time, I've never . . . felt the sinking feeling like I felt when I heard Mayor Bloomberg that day."

I was watching at the same time, and I felt that sinking feeling, too. So, apparently, did the forecasters at NHC.

They don't often communicate directly with local governments—there are too many of them. NHC handles the forecast of the storm itself, and the NWS offices in each locality have the job of interpreting it for the government officials and emergency managers in their areas. But occasionally, when the stakes are high, NHC gets directly involved. It had done so with Katrina; Max Mayfield, NHC director in 2005, called the New Orleans mayor, Ray Nagin, to tell him to evacuate the city. When the situation in New York City appeared to reach a similar level of seriousness, the NHC director in 2012, Rick Knabb, got on the phone as well. At the AMS annual meeting, Knabb said this in his presentation:

"Numerous phone calls with the New York City Office of Emergency Management Saturday evening through Sunday morning. This was a very personal interaction. Myself, Jamie Rhome, our storm surge expert at the Hurricane Center, we were on the phone with New York City all night, into early Sunday morning, culminating with the call on Sunday Morning not long after which they called for evacuations. We don't always talk to every local official, every state official. We're handling the big picture, the WFOs primarily handle the state and local issues. But when we think it's necessary and in this case we certainly did think it was starting Saturday evening, we get directly involved."

16

PREPARATIONS

Decisions

NHC's intervention was successful. By Sunday morning, the mayor's view of the situation had changed. In his morning press conference, he ordered evacuations of Zone A, encompassing the city's lowest-lying areas, home to 375,000 people. An exception to the evacuation order was made for nursing homes and adult homes for those with mental and physical disabilities in Zone A. These would not be evacuated.

Justifying his decision to evacuate, Mayor Bloomberg pointed out that NHC had elevated its storm surge forecast. It was now six to eleven feet, enough to inundate much if not all of Zone A. It had been four to eight feet the night before. These new numbers were not consistently displayed by all official sources; the local Weather Service offices in Mount Holly, New Jersey, and Upton, New York, were still giving lower numbers. But the direct communication with NHC, presumably, had focused the mayor's attention on its new, more severe, and ultimately correct forecast.

The mayor announced that the entire public transportation system (subways, buses, and commuter rail lines) would close at 7:00 that evening, and remain closed all day Monday. He had taken the same step in late August of the previous year, for Hurricane Irene, that being the first time in the history of the city that this was done. This was the second.

This step should have communicated the seriousness of the situation to every New Yorker. While the 375,000 people living in Zone A would be enough to populate a medium-size American city on their own, they were still a small minority of the city's 8 million residents. Many of the rest might have been able to think the storm wouldn't affect them, had the evacuations and the school closings been the extent of it. But New York City depends on its public transportation system completely, profoundly, to an extent that may be hard to grasp for those who live in other cities, in the United States especially, where density is lower and private cars more typical. Without the subways, New York City cannot function. No mayor, this one included, would take lightly a decision to close them.

At the same time, the closing the previous year, during Irene, had been a preventive measure: one that had turned out, in the eyes of many, to be unnecessary. The subways were closed in anticipation that they might flood. But they hadn't flooded. Now, for the second time in two years, some may have viewed this closing as a fire drill rather than a true sign of fire.

Sandy's center was now being forecast, with considerable confidence, to make landfall in central New Jersey. The enormous size of the storm meant that it threatened the coastal regions of that state, New York City, Long Island, and southern New England. The entire region grasped the threat. New Jersey governor Chris Christie had already, on the previous day, ordered evacuations of the state's barrier islands. New York governor Andrew Cuomo now ordered them for New York's, which lie along Long Island's South Shore.

New Jersey Transit, the commuter rail system, began to shut down as well. Like New York's MTA, it began moving its locomotives and passenger cars out of harm's way—except not really. While the MTA moved its rolling stock to safe, high ground, New Jersey Transit moved its cars to low-lying rail yards in Hoboken and Kearny. These yards would flood, and one hundred twenty million dollars' worth of its fleet would be lost.

President Barack Obama declared states of emergency for New York, New Jersey, Maryland, Connecticut, Massachusetts, and the District of Columbia. The Federal Emergency Management Agency (FEMA) had been watching the storm since Wednesday, and had begun sending teams to states along the Eastern Seaboard on Friday, but the presidential declarations allowed FEMA to begin to deploy its resources fully.[42] Mobile Emergency Response Support teams were sent out across the region, bringing in food and water and other emergency supplies and gearing up to respond to the immediate needs that would arise once the storm hit.

These official actions were the start of a swift and strong government response. The president had learned a lesson from the failure of his predecessor, George W. Bush, to take Hurricane Katrina seriously in 2005, and was determined not to repeat that mistake. Bloomberg, too, had been burned by the weather; he was heavily criticized for responding inadequately to a blizzard that buried the city in December 2010.[43] Both leaders were believers in an active, engaged role for government in solving societal problems. This was as opposed to Bush, whose problems in Katrina had started with the hobbling of FEMA's natural disaster response capabilities in favor of counterterrorism after 9/11, when the agency was folded into the new Department of Homeland Security and placed under director Michael Brown.[44]

Chris Christie was closer, politically, to Bush than to Obama or Bloomberg. Not particularly known as an advocate for public solutions to public problems, he had recently torpedoed the construction of a much-needed, federally funded new transit tunnel between his state and New York City. In this instance, though, Christie would put his constituents first and politics a distant second, swinging into gear to work closely and openly with the president.

Persuasion

All those in roles of official responsibility whose jobs included communication with the public (the president, state and local officials, and forecasters at the local NWS offices and NHC) knew that they faced a problem of persuasion. Not everyone would be inclined to obey the evacuation orders. In Cuba, a hurricane evacuation order can be accompanied by soldiers sweeping through neighborhoods to knock on doors and make sure the order is enforced down to the individual. In the United States, a "mandatory" evacuation actually depends on voluntary compliance.

An evacuation is a huge hassle. Many people have nowhere to go but official shelters, which are never going to be an appealing prospect. The roads will be clogged. People may believe they can protect their homes better (either from the storm or from the looters they imagine turning up to take advantage of their absence) by staying in them.

For an evacuation order to be effective, people have to believe they are truly in danger if they stay. Many of those who have never lived through a hurricane will not understand what one can do, and the numbers (forecast wind speeds, storm surges, and rainfall totals) will not mean enough to them.

Perhaps worse than not having lived through a hurricane is having lived through one that didn't live up to the hype. In New York City especially, Irene cast a long shadow over the government's warnings. Both the pitch and quantity of the media stories, and the forecast itself had been similar, in the buildup to Irene, to those of the past week—at least up until today, Sunday, when NHC's surge forecast for Sandy increased.

But Irene hadn't done much damage to the city. The storm's winds had been weaker at landfall than forecast, and the surge had been at the low end of the forecast range: just over four feet, with a forecast of four to eight. The subways hadn't flooded; nor had the streets. Irene had been catastrophic in northern New York and New England, due to rainfall-driven flooding, but the city had gotten off easy. That it had been a very near miss didn't matter to some, who would remember only that a major scare had been followed by a minor storm.

Those in positions to make public statements had to make them convincing enough to overcome not only the nonchalance of those who had never experienced a serious weather disaster, but also post-Irene skepticism. Mayor Bloomberg, in his Sunday press conference, chose to scold: "If you don't evacuate, you're not just putting your own life in danger, you're also endangering lives of our first responders . . . [people who don't evacuate] are not going to get arrested, but they are being, I would argue, very selfish."

On the same day, Sunday, Gary Szatkowski, the meteorologist in charge for the Mount Holly, New Jersey, office of the NWS, included a "personal plea" in his briefing package on Sandy released to the media and officials. In a bulleted list reminiscent of a PowerPoint presentation, he addressed directly those skeptical of the "hype," with a clear and personalized statement of the precautionary principle—we could be wrong, but you don't want to take the chance—and appealed to the memory of the 1962 winter storm, the last time the New Jersey coast had experienced truly major coastal flooding:

- If you are being asked to evacuate a coastal location by state and local officials, please do so.

- If you are reluctant to evacuate, and you know someone who rode out the '62 storm on the barrier islands, ask them if they would do it again.
- If you are still reluctant, think about your loved ones, think about the emergency responders who will be unable to reach you when you make the panicked phone call to be rescued, think about the rescue/recovery teams who will rescue you if you are injured or recover your remains if you do not survive.
- Sandy is an extremely dangerous storm. There will be major property damage, injuries are probably unavoidable, but the goal is zero fatalities.
- If you think the storm is over-hyped and exaggerated, please err on the side of caution. You can call me up on Friday (contact information is at the end of this briefing) and yell at me all you want.
- I will listen to your concerns and comments, but I will tell you in advance, I will be very happy that you are alive & well, no matter how much you yell at me.

Thanks for listening.
Gary Szatkowski—National Weather Service Mount Holly

Graphical Tropical Weather Outlook
National Hurricane Center   Miami, Florida

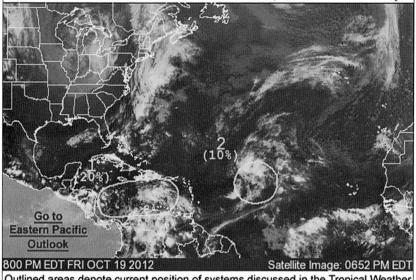

2
(10%)

1
(20%)

Go to
Eastern Pacific
Outlook

800 PM EDT FRI OCT 19 2012   Satellite Image: 0652 PM EDT

Outlined areas denote current position of systems discussed in the Tropical Weather Outlook. Color indicates probability of tropical cyclone formation within 48 hours.

Low <30%   Medium 30-50%   High >50%

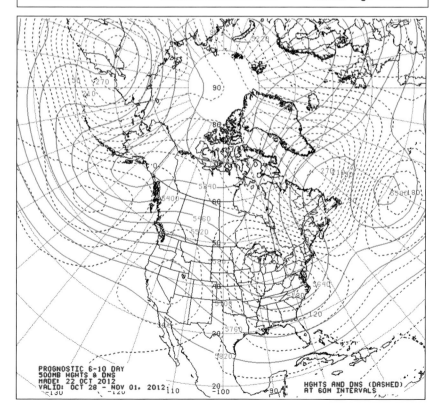

PROGNOSTIC 6-10 DAY
500MB HGHTS & DNS
MADE: 22 OCT 2012
VALID: OCT 28 - NOV 01, 2012

HGHTS AND DNS (DASHED)
AT 60M INTERVALS

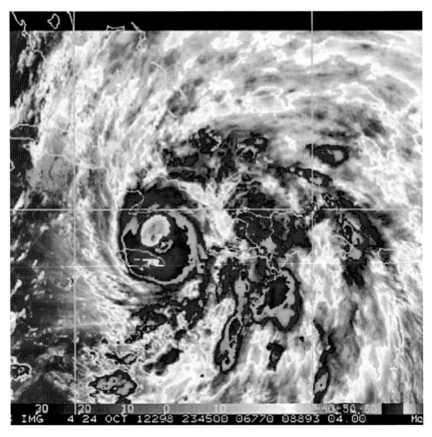

False color infrared satellite image from the GOES-13 satellite at 23:45 UTC on October 24, 2012. The color scale indicates "black body temperature"; low values indicate high, cold cloud tops. (*Courtesy of the Cooperative Institute for Research in the Atmosphere (CIRA), Colorado State University/NOAA*)

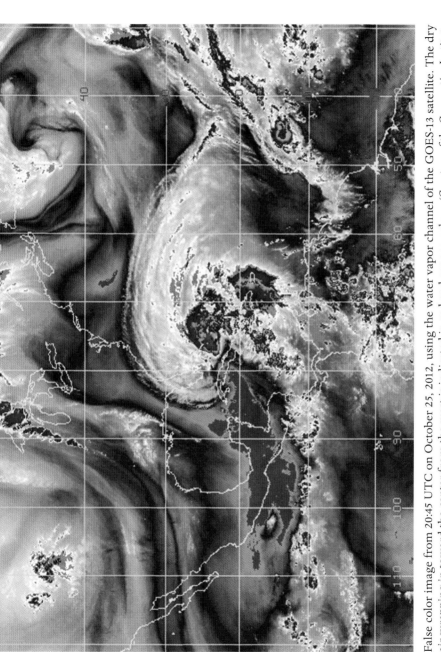

False color image from 20:45 UTC on October 25, 2012, using the water vapor channel of the GOES-13 satellite. The dry air wrapping in toward the center from the west is indicated in red and orange colors. (*Courtesy of the Cooperative Institute for Research in the Atmosphere (CIRA), Colorado State University/NOAA*)

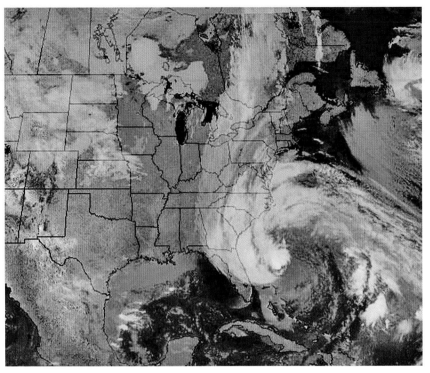

Satellite image composite of infrared images from the GOES-13 and GOES-15 satellites, from approximately 2100 UTC on October 27, 2012. *(Courtesy of the Space Science and Engineering Center (SSEC), University of Wisconsin-Madison, and NOAA)*

Visible image of Sandy from the Moderate Resolution Imaging Spectrometer (MODIS) on NASA's Aqua satellite at 18:20 UTC on October 29, 2012. *(Courtesy of NASA)*

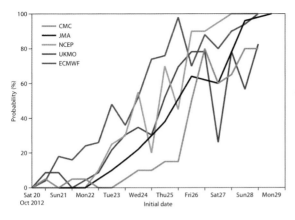

Plot showing the forecast probability (in percentage) of wind speed greater than 38 meters per second at the 850 hectoPascal level (generally indicative of hurricane-force winds at the surface) somewhere inside a radius of 100 kilometers of New York Harbor between 1200 UTC on October 29 and 1200 UTC on October 30, 2012. The different line types indicate the different models: the Canadian Meteorological Centre (CMC), Japan Meteorological Agency (JMA), the United States' National Centers for Environmental Prediction (NCEP), the United Kingdom Met Office (UKMO), and the European Centre for Medium-Range Weather Forecasts (ECMWF). The curves oscillate but steadily rise toward 100 percent, indicating that by the twenty-sixth or twenty-seventh, landfall near New York City had become quite certain. (© 2012 ECMWF, image courtesy Linus Magnusson)

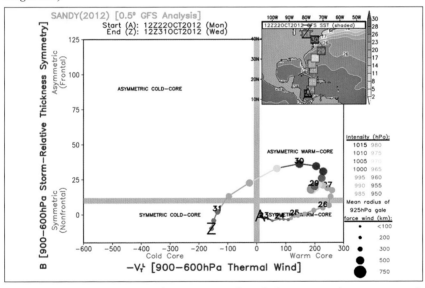

Phase space diagram showing the transition of Sandy from tropical to post-tropical. Horizontal position indicates the relative core temperature (cold-core versus warm-core) while vertical position indicates the degree of symmetry about the center. Lower right is pure tropical while upper left is pure extratropical. The curve traces out Sandy's life in a counterclockwise direction, with a circle each six hours whose size represents Sandy's size and whose color represents its central surface pressure (warmer color means lower pressure). The points "A" and "Z" represent the start and end points on the track, as shown in the insert at upper right, which also marks the dates along the track in boxes color-coded by minimum sea-level pressure, and indicates sea-surface temperature in degrees Centigrade, grayscale at right. (Courtesy of Professor Robert E. Hart, Florida State University)

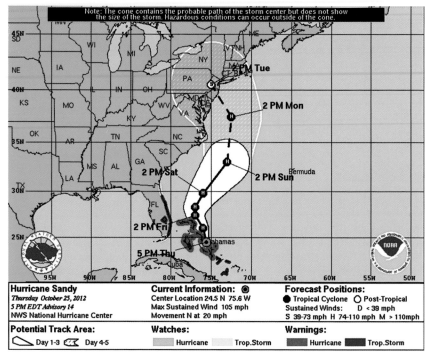

2 PM Tue

2 PM Mon

2 PM Sat

2 PM Sun

Bermuda

2 PM Fri

5 PM Thu

Bahamas

Cuba

Hurricane Sandy	Current Information: ◉	Forecast Positions:
Thursday October 25, 2012	Center Location 24.5 N 75.6 W	● Tropical Cyclone ○ Post-Tropical
5 PM EDT Advisory 14	Max Sustained Wind 105 mph	Sustained Winds: D < 39 mph
NWS National Hurricane Center	Movement N at 20 mph	S 39-73 mph H 74-110 mph M > 110mph

Potential Track Area:	Watches:		Warnings:	
Day 1-3 Day 4-5	Hurricane	Trop.Storm	Hurricane	Trop.Storm

National Hurricane Center forecast from Thursday, October 25, 2012, at 5 p.m. EDT. This forecast was the first that explicitly showed Sandy's landfall as a post-tropical cyclone, indicated by the dashed portion of the track inside the five-day cone of uncertainty.

Flooding of the PATH commuter train station in Hoboken, New Jersey. *(Courtesy of the Port Authority of New York & New Jersey)*

The flooded subway station at South Ferry in lower Manhattan. *(Courtesy of the Metropolitan Transit Authority)*

Taxis in a flooded lot in Hoboken, New Jersey. *(Courtesy of Charles Sykes/Associated Press)*

Breach of the barrier island by the storm tide in Mantoloking, New Jersey. The end of the bridge connecting Mantoloking to the mainland is submerged. *(Courtesy of NOAA)*

Aerial photograph of Manhattan on the night of October 30, 2012. Most of the southern half of Manhattan was without electric power for nearly five days. *(Courtesy of Iwan Baan/ Getty Images)*

Breezy Point, Rockaway Peninsula, Queens, New York, was ravaged by fire during Sandy. *(Courtesy of Mark Lennihan/Associated Press)*

Lines for gasoline in Jamaica, Queens. Gas shortages were widespread for weeks after Sandy. *(Courtesy of Craig Ruttle/Associated Press)*

17

SUNDAY, OCTOBER 28

On Sunday, as the evacuation orders and states of emergency were being issued, the FEMA teams moved into place, and the transit systems prepared to shut down, Sandy continued on its northeastward track. Paralleling the coast, the enormous storm showed no outward indication that it would make landfall in the northeastern United States.

The left turn that Sandy would soon make, having never been made by any Atlantic hurricane on record, would have been impossible to anticipate in previous generations. Had it occurred more than a couple of decades ago, Sandy's landfall would have been difficult to forecast more than a day ahead of time. Had it occurred before 1970, when satellite observations became available, the extent of the threat might have been unforeseen until it materialized. As bad as Sandy would prove to be, it could have been enormously worse, like the hurricane that destroyed Galveston, Texas, with no warning, in 1900; or, much closer to home, the fast-moving 1938 storm whose devastation of Long Island and New England also came as a surprise—but orders of magnitude more harmful for being visited on a much larger and more heavily populated area.

In 2012, though, by Sunday morning, October 28, thirty-six hours before landfall, the forecast had been consistent and confident for several days. The advisories and discussions posted by NHC on Sunday, read after those from the week previous, are repetitive and anticlimactic, despite the severity of the events they predict. The major change was the substantial uptick in the storm surge forecast, from four to eight feet to the new six to eleven feet, which had just convinced the Bloomberg administration to ramp up its actions dramatically. There was little other drama in the forecasters' words, although those words spelled out very accurately the meteorological facts that would cause the disaster.

The advisories of 11:00 a.m. and 5:00 p.m. Eastern Daylight Time on Sunday reported the current facts about Sandy: maximum winds of sixty-five knots, still hurricane strength, and a central surface pressure just above 950 hectoPascals, astonishingly low for a storm reaching the Atlantic midlatitudes. They predicted that the storm would make its left turn late Sunday night or Monday morning, and make landfall Monday night. They predicted that hurricane-force winds would influence sections of coast from Virginia to Massachusetts, and tropical storm–force winds a still larger area, though, for the reasons they had explained in previous advisories, they didn't issue tropical storm or hurricane warnings or watches for any of these places. They explained that the storm would change from one species to another, but it wouldn't matter, as explained by forecaster Stacy Stewart in the 5:00 p.m. advisory:

SANDY IS EXPECTED TO TRANSITION INTO A FRONTAL OR WINTERTIME LOW PRESSURE SYSTEM PRIOR TO LANDFALL. HOWEVER . . . THIS TRANSITION WILL NOT BE ACCOMPANIED BY A WEAKENING OF THE SYSTEM . . . AND IN FACT A LITTLE STRENGTHENING IS POSSIBLE DURING THIS PROCESS.

After stating the storm surge forecast numbers, Sunday's advisories repeated the verbal statements from earlier ones about the potential severity and duration of the coastal flooding, and the reminder that "these conditions will occur regardless of whether Sandy is a tropical or post-tropical cyclone."

Near the end of the 5:00 p.m. advisory came a prediction that had not been in any of the previous ones:

SNOWFALL . . . SNOW ACCUMULATIONS OF 2 TO 3 FEET ARE EXPECTED IN THE MOUNTAINS OF WEST VIRGINIA . . . WITH LOCALLY HIGHER TOTALS TONIGHT THROUGH TUESDAY NIGHT.

SNOWFALL OF 1 TO 2 FEET IS EXPECTED IN THE MOUNTAINS OF
SOUTHWESTERN VIRGINIA TO THE KENTUCKY BORDER . . . WITH
12 TO 18 INCHES OF SNOW POSSIBLE IN THE MOUNTAINS NEAR
THE NORTH CAROLINA/TENNESSEE BORDER.

Snow in an advisory from the National Hurricane Center—this had
never happened before, and by itself would have indicated, if nothing else
had, how truly exceptional this storm was. Not just snow, but a huge quantity, in October, coming with a hurricane.

Satellite images at this time show cloud cover over the entire northeastern United States, eastern Canada, and the North Atlantic.[45] The cloud becomes thick as it hooks from the northwest into the center, at this time
offshore of the Carolinas, still quite far south of its eventual landfall point.
The eastern side of the hurricane is much more exposed, the clouds thinner,
but clearly outlining the spiraling circulation. Just north-northwest of the
center, a particularly intense burst of convection is evident in a dense, thick
blob of cloud.

Convective bursts had been flaring in this spot, relative to the center,
for days. These bursts justified the continued identification of the storm
as tropical. With the low-level winds on the large scale blowing from the
northeast, and the upper-level winds blowing from the southwest, associated with the strong cold front approaching the coast, the shear vector (the
difference between the upper- and lower-level winds, expressed as a vector,
with both speed and direction) pointed approximately from southwest to
northeast. This placed the repeated convective bursts to the left of the shear
vector. This "downshear left" quadrant is where the strongest convection is
typically seen in the inner core of a hurricane exposed to significant wind
shear, but surviving it.[46]

At the same time, the enormous cloud shield to the west formed the
comma shape typical of large, powerful extratropical cyclones. Sandy's size,
power, and hybrid character are all apparent from Sunday's satellite images.

Meanwhile, the water had begun to rise. By Sunday evening, the tide
gauge at the Battery showed a level two feet above the normal tide. Two feet
poses no danger, but that was just the current stage of a slow increase that
had become clearly detectable by about Saturday morning.

18

MONDAY, OCTOBER 29

By Monday morning, October 29, the sky in New York City had turned dark. Still, in Morningside Heights, high above the Hudson on northern Manhattan's West Side, one couldn't have grasped the developing disaster simply by stepping outside. Little rain fell. The winds were powerful enough to be intimidating. They were not, in the city, at hurricane strength, but over the course of the day, they would bring down many, many trees.

Miscellaneous other urban items, some of them very large, would also fall. At 2:30 p.m., an 850-foot construction crane on West Fifty-Seventh Street, in the heart of Midtown, folded over, metal crunching.[47] Collapsed, it would not, however, fall, but would stay poised over the center of Manhattan for days.

For the city, though, its power lines mostly belowground, its buildings mostly concrete and steel, the wind's direct effects would be a footnote to the storm's real impact.

New Yorkers had been sternly instructed by the mayor to stay inside for their own safety. I had been inside most of the weekend already. I had been talking to the media—Brian Lehrer, at WNYC radio, was doing hours of special storm coverage, and called me repeatedly for brief on-air conversations—and watching the data online: the satellite images, the model output, the forecasts, the tide gauges. By mid-morning, the Battery gauge was showing up to four feet above the normal tide. At the first high tide of the day, that put the total storm tide at nine feet above the low-tide mark. That was about how high it had gotten in Irene, the year before. This time, it was just a warm-up.

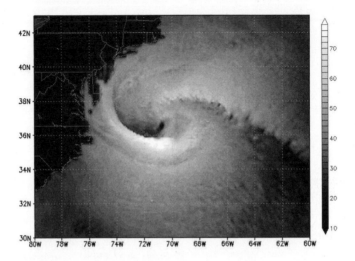

Surface wind speed map, in knots (gray scale at *right*), from a twelve-hour forecast simulation with the Hurricane Weather Research and Forecast model, made in real time at 00 UTC, October 29, 2012, showing Sandy's winds at 1200 UTC on that day, about twelve hours before landfall. (© 2014 Fuqing Zhang and Yonghui Weng)

I watched Sandy mostly through my computer screen, from a distance, as a meteorological entity reflected in maps and graphs. News reports of the damage were beginning to flood the airwaves and the Internet; I didn't look at them. I did grasp the scope of what was happening to the city and the region, but I couldn't take it in. I was focused on the natural phenomenon, the storm itself. I was overstimulated, excited, and afraid, all at once. I knew I wouldn't experience the worst of the storm personally, because I live on a hill, far too high for the surge to reach. I did think it pretty likely that we would lose power, Internet, and cell phone service. We never did.

At about ten in the evening, I couldn't stand it. This was the storm of my lifetime, and I was seeing it just in electronic form, and through our windows onto West 113th Street, the buildings close across the street allowing a view of only a narrow slice of sky. The surge had just peaked.

The Hudson River was just down the hill. I knew I was never likely to see the water that high if I didn't see it now. I also knew that our mayor's warnings to stay inside were not just an idle precaution; one should not go out in a hurricane.

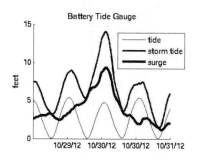

Water level at the Battery, at the southern tip of Manhattan, as measured by the NOAA tide gauge, as a function of time between 00 UTC on October 29 and 00 UTC on October 31, 2012. (Eastern Daylight Time was four hours behind UTC.) The thin curve shows the astronomical tide, the thickest curve shows the storm surge, and the curve with intermediate thickness shows the storm tide, which is the sum of the other two. *(Data from NOAA)*

Still, I had to go downstairs and, at least, stand in our doorway for a few minutes. I needed to be outside just a little. I promised my wife I would come back up . . . unless it looked really safe outside. In that case, maybe I would walk down to the river.

When I got downstairs, no more than a drizzle was falling. The wind was gusty but not severe. I walked out and north through our neighborhood, Morningside Heights. Some large trees had fallen, and one or two chunks of concrete lay shattered on the streets. Poles holding street signs listed. It was clear something had happened, but nothing near a catastrophe—not for my neighborhood. The wind wasn't strong at the ground level, but was gusty in the branches higher up. I looked up constantly, highly conscious that large branches falling were the only real danger to me. I walked north on Claremont Avenue, and then downhill on Riverside Drive, aiming for Manhattanville, the region hard by the river and just north of Riverside Park, where you can get to the Hudson directly from the street. I passed a handful of other people also out walking, all calm.

I arrived down at 12th Avenue and 133rd street. The Henry Hudson

Parkway is elevated here, above the street, and above the parking lot for Fairway supermarket. Just east of the parkway is the access road for it, at street level, and then, bordering the river, the West Harlem Waterfront Park. The park and access road were all underwater. The river had come up over them, over the parking lot, up under the parkway. The water was still.

It was 11:00 p.m. Three hours earlier, Sandy's center, with a surface pressure of 946 hectoPascals, the lowest ever recorded in the northeastern United States, had made landfall near Atlantic City.

19

THE DAMAGE

Manhattan

Lower Manhattan is fringed by land that, historically, was either wetland or open water: the Hudson and East Rivers. Since colonial times, the water has been filled in, the wetlands made solid, landfills built to extend the island into the rivers and the harbor, and all of it built on.[48] But all that land still lies low.

On the evening of Monday, October 29, Sandy drove the water back over all of it: the Financial District; Battery Park City; eastern parts of Chinatown, and Alphabet City, in the far East Village; and, farther north, sections of Chelsea and Gramercy, west and east. This is among the most densely inhabited, heavily used, and economically active real estate in the developed world.

Manhattan Island, New York City's economic center, has a dense web of connections to the land masses across the water. Many of those are underground. From Lower and Midtown Manhattan, tunnels go east to Brooklyn and Queens, and west to New Jersey. Some carry automobiles; some carry trains.

The entrances to the tunnels are close to sea level. The authorities tried to keep the water out of them. The MTA had, following the system shutdown the previous evening, raced to cover sidewalk air vents and station entrances. They attached twelve hundred sheets of plywood and heaped fifteen thousand sandbags.[49] The impromptu plywood dam they threw up to protect the 148th Street tunnel in East Harlem worked, saving the No. 3 line from flooding.[50] The rest of their barriers didn't work. Seven subway tunnels flooded—essentially all those going east to Brooklyn and Queens. Seawater also filled the PATH tunnels, which carry commuters from New Jersey. The automobile tunnels did no better. The Brooklyn–Battery Tunnel, Midtown Tunnel to Queens, and Holland Tunnel to New Jersey all flooded; of the automobile tunnels in and out of Manhattan, only the Lincoln Tunnel stayed dry.

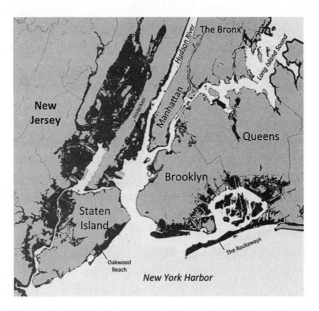

Map showing the extent of the flooding in New York City and the nearest part of New Jersey. The lightest shade indicates water, the next lightest is dry land, and the darkest indicates land that was inundated during Sandy. *(Courtesy of Marit Larson; data from FEMA)*

Manhattan's power lines are underground, so they couldn't come down in the wind. But transformers at the lowest-lying of Con Edison's Manhattan substations were prepared for storm tides of ten to twelve feet. Measured from the mean low tide, Sandy's was fourteen feet. The East River, Thirteenth Street, and Seaport substations in southern Manhattan flooded, and failed. Some of Lower Manhattan's power was already out, preemptively shut down by Con Ed to protect the equipment (fully expected to be flooded) from being destroyed by shorting out. Now most of the rest of Lower Manhattan, and much of Midtown, lay dark. The blackout went from the island's tip up to Thirty-Ninth Street.

At 8:12 p.m. on Monday, just before the power went out at the East River and Thirteenth Street substations, some "lower-voltage equipment associated with a single East Thirteenth Street substation high-voltage transformer failed catastrophically,"[51] as saltwater shorted it out. From across the East River, someone with a video camera saw this happen and posted the file to YouTube.

A bright glow suddenly flares at ground level amid a placid, distant Manhattan skyline. It brightens, then suddenly brightens much more, shockingly so, saturating the image. Over a minute, this repeats several times. Con Ed engineers, in a report presented at a conference the following spring, wrote that "The failure caused a dramatic arcing fault, and looked and sounded like an explosion." Yes, it did. They go on to point out that this failure alone, contrary to popular belief, would not have been enough to shut down the entire station, let alone all of Lower Manhattan's power. It was just the failure that everyone saw. Nonetheless, within the hour, half the island was dark.

Brooklyn, Queens, Long Island

Like Manhattan, Brooklyn and Queens have coastal edges that are reclaimed swamp, landfill, or otherwise low-lying. All went under: Red Hook and the Navy Yard on the East River; Coney Island, Brighton Beach, and the other areas along Jamaica Bay on the south. In northern Queens, and in the Bronx across Long Island Sound, many small patches flooded, but not as badly as they could have; the tides in the Sound are around three hours behind those in the harbor, so the peak surge didn't match the Sound's high tide, and the total storm tide wasn't as high. But the runways of LaGuardia Airport flooded, as did those of JFK in the south, though the terminal buildings of both were unharmed.

Brooklyn and Queens also have barrier islands, the western edge of the thin chain that stretches eastward off Long Island's southern shore. From the Rockaways to Long Beach to Fire Island to Westhampton—all were submerged.

The barrier islands are studies in socioeconomic contrasts. Fire Island, off the South Shore of central Long Island, is accessible only by ferry and has mostly vacation homes and almost no automobiles on it. Most of the island is undeveloped "national seashore," managed by the National Parks Service. The east end of the Rockaways has housing projects. At the Rockaways' west end, the private community of Breezy Point is an island within an island. Its residents, some of them there year-round but many, as on Fire Island, summer vacationers, live in beach bungalows that have been in families for generations.

All of the Rockaways was in Zone A. As elsewhere, some evacuated, but not everyone. Some didn't evacuate who had done so with Irene, made complacent by that previous storm's mildness.[52]

In Breezy Point and Rockaway Beach, some were still there when saltwater reached electrical equipment in flooded homes. The local utilities, the Long Island Power Authority and National Grid, had left the power on, contrary to the terms of the evacuation order.[53] With the fire department hobbled by the flooding, the fires that broke out spread, incinerating neighborhoods. More than one hundred homes in Breezy Point burned, all of them within a contiguous area. Aerial images afterward show surviving houses on two sides of a charred abscess, little left above the foundations.

Staten Island

As Sandy's center came ashore near Atlantic City, its winds blew easterly north of there, driving the surge onshore. The New Jersey and Long Island shores form a wedge. The wedge compressed the incoming surge like a funnel, amplifying it until it reached the triangle's point. Staten Island lies at that point.

The city's least urban borough, Staten Island has many neighborhoods, particularly along its southeast-facing shore, made up of single-family homes. These neighborhoods face not just the harbor, but also the Atlantic. Standing on South Beach today, one sees Brooklyn and the Rockaways across the water on the left, clusters of tall buildings along the shore, all connected to Staten Island's north by the Verrazano-Narrows Bridge. New Jersey lies on the right, the highlands sloping down to flat Sandy Hook, barely visible but for the trees raising its profile above the water, enclosing the harbor's south. Straight out from shore, one sees Hoffman Island, a small, artificial patch of land built in the nineteenth century. Behind that, one is looking at the open sea. Facing southeast, one's line of sight would lead, unimpeded by anything but the curvature of the earth, to Dakar, Senegal, where African easterly waves, some of them future Caribbean hurricanes, leave the continent.

The neighborhoods on this shore (Ocean Breeze, South Beach, Midland Beach, Oakwood Beach) were built on swampland, behind low dunes. Father Capodanno Boulevard, a wide artery, runs inland of the parks and boardwalk lining the shore. Perhaps a hundred yards from the beach, the road is separated from it by flat parking lots and ball fields. Behind the boulevard, although one appears to be about level with the ocean as one looks east, the land slopes *down* as one goes west.

In many places, the wetlands are still there. But there are streets and houses, too, some of their roofs just barely clearing the *Phragmites*.

These neighborhoods flooded, catastrophically.

Right on the shore, the high water wasn't still, as where I was viewing it, along the Hudson in northern Manhattan. Here, facing the expanse of the Atlantic, with the powerful easterlies on Sandy's north flank driving onshore with a fetch thousands of miles across, gigantic waves brought the sea in with tremendous force.

In Oakwood Beach, the sea was lined by ten or twenty cottages, marshland behind them. Remnants of the day when the neighborhood was a summer beach resort, these had become permanent residences. After Sandy, only two survived in place. Some were destroyed entirely. Others were torn loose and swept far from their foundations. In aerial photos taken shortly afterward, one sees an isolated house in the middle of an expanse of green, a skid mark several hundred yards long showing the track taken by the house to its resting place.[54]

Boats, too, were sent inland, from the marina at Great Kills Park (itself destroyed) and other smaller docks; one sixty-two-foot yacht was later found on the roof of a restaurant.[55]

The water reached Hylan Boulevard, the main drag that parallels the shore. Hylan runs close to a mile inland north of Great Kills, delineating the inland extent of Oakwood, Midland Beach, and South Beach. The parts of these neighborhoods farther back from the shore didn't face the waves directly, but they went deep underwater.

Sandy killed twenty-three people in Staten Island, the lion's share of the forty-three who died in New York City.[56] Most drowned; a few were crushed by debris or electrocuted.[57] Some died in their homes; others in cars, trying to flee. Two young boys, Brendan and Connor Moore, two and four years old, were swept away as their mother took them out of their flooded car, trying to get them to safety. Their bodies were found days later. A long search ended when a Parks Department employee (from the Natural Resources Group, which works on monitoring and restoring Staten Island's wetlands) guided the police, up to that point unable to find the two lost children on their own, through the wetlands behind Father Capodanno Boulevard and to the boys' bodies.

Photo from Ocean Breeze, Staten Island, one of the worst-hit neighborhoods. This photo, taken in November 2013, shows the level reached by the water during Sandy, as marked by residents on the pole near the top of the image. The author is shown for scale. As the sign indicates, some residents in Ocean Breeze had not yet been able to return to their homes at the time of the photo, a year after Sandy's landfall. *(Courtesy of Marit Larson)*

New Jersey

New Jersey's coastline is mostly barrier island, narrow strips of sand dune, separated from the mainland when the end of the ice age raised the sea and flooded the lowland behind it, making Barnegat Bay and Little Egg Harbor.

Island Beach, the northern of the two large barrier islands, runs from Bay Head in the north (the end of the New Jersey Transit line to the city) down to Island Beach State Park in the south. In between ran very different types of towns, from large, Hamptons-style oceanfront houses in Mantoloking to small cottages in tightly packed rows perpendicular to the ocean in Camp Osborn. In many places, the island is narrow enough that from Ocean Boulevard, essentially the one main street running the length of it, one can easily see both the Atlantic to the east and Barnegat Bay to the west just by turning one's head.

The surge along the New Jersey Shore, lacking the funneling effect of the harbor to the north, may have been slightly lower than in the city. But the tide, close to that in the city, was similarly high at the peak surge. The barrier islands are flat, and the waves came in hard.

In Mantoloking, the island breached. The sea split Island Beach in two right by the bridge to the mainland; multimillion-dollar beachfront houses around it were swept away like sand castles. The island end of the bridge was submerged; the next bridge to the south, in Seaside Park, was rendered impassable by downed telephone poles,[58] cutting off all access to the mainland for anyone remaining on the south side of the Mantoloking Bridge: in Ortley Beach, Lavallette, Seaside Park.

Seaside Heights is famous for its boardwalk, a dense strip of shops selling funnel cakes, hot dogs, T-shirts with garish slogans. The amusement park is its centerpiece. The boardwalk was destroyed. The collapsing pier dropped the roller coaster into the Atlantic. As gas lines ruptured, Camp Osborn, a small community next to Seaside Heights settled in the 1920s, caught fire in the midst of the flood and burned to the ground.

Farther north, Hoboken, New Jersey, lies across the Hudson from Lower Manhattan's west side. From a historically unglamorous past, high real estate prices in New York City and easy access to Manhattan via the PATH train had gentrified it since the 1980s, making it feel like an outpost of hipster Brooklyn. But much of Hoboken lies below sea level. Many residents were even lower, as high density had made basement apartments economically viable. More than half the city flooded, either from the surge or from the sewers backing up.[59] The water pouring into the Hoboken PATH station became an iconic image of the storm. Service would not be restored for three months.

More than thirty people died in Sandy in New Jersey. Hundreds of thousands of homes in the state were damaged or destroyed; tens of thousands were rendered uninhabitable.[60]

20

THE TOLL

According to a report by the Centers for Disease Control and Prevention, 117 people died in the United States because of Sandy. Of those, 53 were in New York State and 34 in New Jersey, with the rest in Pennsylvania, Maryland, West Virginia, and Connecticut. Forty people drowned (32 of those in New York) and another 19 were killed by trauma from being "crushed, cut or struck." Thirty-eight deaths were categorized as "indirect." Of those, the most common cause was poisoning by carbon monoxide, presumably emitted from portable generators and heaters.[61]

This death toll is almost certainly conservative. The CDC report is based on data from the American Red Cross gathered in the first month after the storm, from October 28 to November 30 of 2012. But the harm inflicted by Sandy was not done then.

The evacuation order issued for Zone A in New York City specifically excluded nursing homes and adult care facilities. The evacuation of "elderly and vulnerable people" was thought to be too risky.[62] This turned out to be a severely unfortunate miscalculation.

Many facilities flooded. They lost power, water, heat, and food. Under these conditions, they had no choice but to evacuate. But now they were doing so under conditions much more challenging than they would have been before the storm. Elevators couldn't move without power, and patients had to be carried downstairs. Transportation and communication were difficult.

Many patients then endured long waits to be transported far away, in some cases to facilities that lacked information critical to their care.

One nursing home in Coney Island reported that 125 of its residents died in the six months after Sandy, out of 611 who were living there before.[63] While it may be impossible to attribute all those deaths conclusively to the "transfer trauma" associated with the storm, 125 out of 611 is a very large percentage to lose in a six-month period. If even a significant fraction of these deaths was caused by Sandy, and if other facilities suffered comparable losses, then the death toll related to Sandy is much higher than the CDC reported.

This problem is not unique to Sandy. It is often difficult to determine the number of deaths attributable to a natural disaster, and different sources often give conflicting numbers. Accounting for indirect deaths is particularly difficult. Official estimates are most likely to be low. My colleague John Mutter and his associate Richard Garfield did a study of deaths caused indirectly by Hurricane Katrina. They concluded, in 2008, that "more than two years after the event we do not yet know how many people died as a result of Hurricane Katrina,"[64] but that the total could be as much as twice the official figure, with most of the additional deaths among the poor or elderly.

Sandy's official U.S. death toll, 117, is much lower than Katrina's, at 1,833.[65] Even if the number of unofficial, indirect deaths were to multiply Sandy's total by two (or even three or four), it would remain much lower. Not so the economic damage.

NHC's report estimated the economic damage from Sandy at fifty billion dollars. Other estimates have come in higher; the reinsurance company Aon Benfield's number is sixty-five billion.[66] The U.S. Congress, after an ugly ninety-one-day period of delay driven by partisan politics, allocated fifty-one billion dollars for relief to the states directly affected by the storm. This was much needed and appreciated, but smaller than the eighty-two billion dollars requested by the governors of New York, New Jersey, and Connecticut.[67] Estimates of economic damage from past storms clearly put Katrina on top, with numbers on the order of a hundred billion dollars or more.[68] Before Sandy, Hurricane Andrew in 1992 was second, at forty-four billion in today's dollars. Sandy seems almost certain to top that, making it the second-worst hurricane disaster in U.S. history from an economic point of view.

The scholar Roger Pielke Jr. keeps track of hurricane economic damages and "normalizes" them to account not just for inflation but also economic development. If past storms had hit the same places today as they did when they happened, the damage would almost inevitably be worse, since more people and structures are in those places now. Accounting for this, a fifty-billion-dollar total would place Sandy seventh on Pielke's list,[69] below not just Andrew and Katrina, but also *two* Galveston hurricanes (the famous 1900 one, but also the 1915 storm) and, at the top, the "Great Miami Hurricane" of 1926. A sixty-five-billion-dollar total would bring Sandy just above Andrew.

Yet while storms in the past might have done more damage had they hit today, they didn't. It is important to understand the role of economic development in natural disasters—and we'll come back to this. But Sandy happened when and where it did. History will record it as a high-water mark, literally and figuratively.

21

IMMEDIATE AFTERMATH

By the morning of Tuesday, October 30, the surge had receded, but the mid-Atlantic coastal United States woke up wrecked.

The worst hit, in the low-lying flooded coastal areas, faced destroyed homes, businesses, and, in some cases, lives. In the short term, they needed relief: food, water, shelter, medical care. In Staten Island's eastern shore, the Rockaways, Red Hook, along the New Jersey Shore, and all the other places directly hit by the surge, the structures of modern life were literally gone.

The relief effort would come from many quarters. From FEMA, from the states, from the New York City Office of Emergency Management, but also from a great number of volunteers rushing in, some organized, some not. These efforts would relieve some of the immediate suffering. The reconstruction effort would be much more protracted. In many areas, it is not over now, more than a year later as I write. Some displaced by Sandy are still not back in their homes; some will never be. That is a story for another book.

But in the immediate aftermath, the damage extended far wider than the water had.

Power was out to 1.1 million people in New York City alone. In the region as a whole, more than 8.5 million people were without electricity at the peak of the outages.[70] Sandy's winds, in most places below hurricane strength, had not been strong enough to destroy many buildings, but they had brought down countless trees, and overhead power lines with them, across the suburbs of New Jersey, New York, Connecticut, and the adjacent states. For some, the lights would come back on soon; for many, it would take much longer.

Nighttime video shot during the first week by Dana O'Keefe and Alex Kliment[71] shows dark Manhattan streets with cars and buses going by, navigating without traffic signals, their own lights on, incongruously. The scene is spooky for the simultaneity of its normalcy and utter extremity.

Capt. Jonathan Boulware, interim director of the South Street Seaport Museum on Lower Manhattan's East River waterfront, described the neighborhood in those first days as "Mad Max," with the lack of functioning infrastructure and the temporary infrastructure being brought in: tubes draped everywhere on the streets, generators running their engines loudly. He described the feeling when leaving the area to stay with a friend on the north end of Central Park (blocks from where I live), where everything but public transit was working, as having moved from the third world to the first just by crossing Midtown.

Down by Boulware's museum, the surge had reached, poetically, to Water Street; the street is named that because, in the time of the Dutch settlement, it was the shore. The blocks east of Water Street were built on landfill, of Dutch and then English construction, centuries earlier. Reclaimed by water when Sandy hit, they had belonged to the water originally.

The dark half of Manhattan would begin lighting up again on Friday, November 2, and come back on completely on Saturday. The crime waves that some had feared did not materialize.

The subways, similarly, would come back, mostly, in about a week. Like the blackout, the duration of the subway shutdown, too, was totally unprecedented; the system had never been disabled for anywhere near this long since its construction.

In the first weeks after the storm, as is inevitably the case with truly major disasters, systems on which our modern developed society depends experienced continued cascading failures in the New York metropolitan area. The failures demonstrated all these systems' interconnectedness, complexity, and vulnerability.

Gasoline shortages developed. The immediate cause was the disruption of supply lines from ports. In some places, particularly outside the city, the situation was made worse by the loss of power to gas stations, most of them lacking backup generators. Even if they had gasoline, they couldn't pump it.

Real shortages were made worse by imagined ones, as people rushed to fill up, fearing that they wouldn't be able to do so later. Governor Christie ordered odd/even-day rationing in New Jersey first; in New York City, Mayor Bloomberg resisted doing the same until Thursday, November 8, ten days after Sandy's landfall.[72]

Seven years earlier, Hurricane Katrina's devastation of New Orleans had met with a government response (local, state, and above all federal) that had been at best deeply inadequate, at worst counterproductive. The response to Sandy was radically different. President Obama, aware of his predecessor's failure in Katrina and facing an election in a week, promised on Tuesday, the day after the storm, that there would be "no red tape"—all layers of government should cooperate in a sustained response, facilitated at the federal level. While there would be shortcomings, and valid complaints, in the following weeks and months, the government's performance in the aftermath of Sandy would prove to be a dramatic improvement on that in Katrina.

FEMA, the agency that had performed perhaps worst of all in Katrina, shifted into high gear quickly with Sandy. The FEMA operations center, its Joint Field Office (JFO) in New York—there would be one in each state affected by the storm—would quickly grow to employ several thousand staff, many reservists called up from around the country. They worked first on providing food, water, and other critical supplies; on finding housing for those who could not return to their homes, first short term and then longer; and eventually on helping people process claims for financial assistance.

Michael Byrne, the head of the New York JFO, came to speak to my class in the spring of 2013.[73] Byrne is a former New York City firefighter who grew up in an East Harlem housing project, then was involved in the founding of the City's Office of Emergency Management before moving to FEMA. He left the agency during the Bush years—he was not there for Katrina—but later came back. Byrne is one of the few within FEMA who can lead the response to a Type 1 disaster, the worst kind. He had been in charge after 9/11, the spring 2011 tornado outbreaks, and Hurricane Isaac.

Byrne described the unique challenges of working in our dense urban environment and some of the unusual dimensions of the response. In the first days, as his team moved to secure food and water, they realized that, among other things, they would need to order many kosher meals, and did so. Finding shelter for displaced people, in the days and weeks after landfall,

also posed a unique problem in a metropolitan area with almost no undeveloped land and little vacant housing. Byrne said FEMA "could have taken
over Central Park" for tent cities, but opted not to go that route in real-
estate-challenged New York City. The agency took a new approach where
possible, repairing damaged homes just enough to allow people back in,
rather than trying to find them temporary housing for long enough to allow
more complete repairs to those homes.

Byrne described the "velocity" at which new problems appeared, all
needing to be solved with inadequate information, and the sheer magnitude
of the operation. Sixteen thousand federal employees would be involved in
the response, thirty-eight hundred of those employed by FEMA. For the first
time, the U.S. Marines landed on domestic beaches (in Staten Island and the
Rockaways) in response to a disaster.

Despite the magnitude of the government efforts, and the tremendous
improvement they represented over Katrina seven years earlier, help did not
reach many in the hardest-hit areas in the early days. In the neighborhoods
that suffered the most damage, the most effective first responders were volunteers. The primary entity organizing those volunteers, the most effective
and widespread, was a uniquely New York product, one whose time in the
news had been thought over.

Occupy

On September 17 of the previous year, 2011, a group of protesters moved
into Zuccotti Park, a city block–size area in Manhattan's Financial District.
Angry at the financial industry for its role in the economic crisis, at government for its failure to hold Wall Street accountable, and at the overall
deepening inequality in the country, they had decided that the way to focus
attention on these issues was simply to start living in the park. Over the
weeks that followed, a quasi-organization would form. Though unfocused,
prioritizing the purity of internal process (strict democratic principles and
consensus decision making) over specific demands, Occupy Wall Street generated an energy not seen in the American left in many years.

The protesters left Zuccotti Park after two months, forced out by the
police on November 15. The protests had long since spread beyond New
York City, with Occupy branches emerging around the world. Still, by the
time Sandy struck, the movement seemed to many to have become history

already. When the waters flooded the South Street Seaport Museum, one of the exhibits currently on display was on Occupy Wall Street. (The museum, its basement utilities not yet repaired, had not reopened when I visited in late 2013; the signs for the Occupy exhibit were still up in the entrance hall.)

But the momentum had not dissipated entirely. On the last weekend of October 2012, as Sandy was moving north, about twenty-five core members of Occupy Wall Street were at a retreat in upstate New York reflecting on the time since leaving the park and trying to decide what to do next. They saw the forecasts but didn't talk much about the storm.

As the Occupiers living in New York City went home, they were thinking about how they could help if the storm's impact turned out to be as bad as the forecasts were suggesting it would be. Occupy had formed networks of people, mostly in their twenties and thirties, many well educated, some with broader involvement in activism or community organizing, all tech-savvy and effective at digital communication. Without any coordination at first, they began doing things. Some knew how to raise funds, and now set up Facebook and WePay pages to coordinate donations for storm victims. Another set up a Sandy page on the Recovers site, collecting hotline numbers. By the thirtieth, the day after landfall, connections between these and other initiatives began forming to constitute an amorphous network.

That day, October 30, a few Occupiers stopped by the Red Hook Initiative, a community center in Red Hook, Brooklyn, one of the lowest-lying and hardest-hit neighborhoods in the city, to see what they could do to help. By the next day, the thirty-first, they had helped to set up a drop-off location at a church in nearby Sunset Park, and organized the first conference call for what quickly was to become Occupy Sandy.

Working with local groups, Occupy set up relief centers quickly in Red Hook and the Rockaways, and not long after in Coney Island, Staten Island, Sheepshead Bay, and New Jersey. The centers took in donations of money and supplies, organized volunteers, and mobilized those resources to go where they were needed, often in places where no government relief workers or those from any other previously established entity such as the Red Cross were to be found. Occupy estimates that it organized forty thousand volunteers through its website and served two hundred thousand meals.

The experience and motivation of the Occupiers proved invaluable. Fast communication by text message, conference calls, and social media allowed efficient deployment of people and supplies to where they were most needed. The experience gained in Zuccotti Park feeding and clothing people in an unfavorable environment was exactly appropriate. Tamara Shapiro, one of the core group, described the spontaneity of the effort: there had been no prior communication, no planning, just "people doing what they knew how to do."

Occupy's focus on inequality was also relevant. Just as Sandy's impacts in the Caribbean had been disproportionately bad in destitute, unprepared Haiti, its impacts within New York City were felt disproportionately by the most vulnerable. Nursing homes in Zone A had not been evacuated, leaving many residents trapped. The disabled, in many cases, could not comply with the evacuation orders in time. Many New York City public housing projects are tall buildings; without power, residents on high floors could not leave if they were too infirm to descend many flights of stairs. Occupy volunteers reached many of these places first. A city press release of November 9 states that mobile medical teams were being dispatched to go "door-to-door in tall residential buildings."[74] That was eleven days after the storm. It was fortunate that volunteers, many of them organized by Occupy, had already been there.

Politics

In the first days and weeks after the storm, science met politics in ways as unprecedented as Sandy itself.

Homeowners surveyed the damage, hoping to make insurance claims. In both New York and New Jersey, insurance policies covering wind damage treated claims differently depending on whether the storm was a hurricane. If the storm was a hurricane, the policyholder would have to pay a deductible; otherwise, not. The trigger was not the wind speed itself (as one might have expected it to be, since that is what determines the actual damage), but whether NHC had called the storm a hurricane. (This remains true post-Sandy.[75]) A "post tropical" storm with hurricane-strength winds was a better scenario for homeowners and a worse one for insurance companies than a hurricane. The same technicality that had confounded the forecast and warning process lived on to confound the recovery.

Although NHC had declared Sandy post-tropical at landfall, it had clearly been a close call. While that call couldn't be changed for the purpose of the forecasts and warnings (already in the past), it wasn't yet final for other purposes, including insurance. After the hurricane season ends, NHC routinely reviews all storms, producing a set of final reports that can, in some instances, change judgments made in the heat of the forecast moment. New data can become available, or the same data can be interpreted differently given more time. With the report on Sandy not to be issued for several months, the question of tropical versus post-tropical, and thus deductible versus no deductible, was not settled.

In the first weeks, area politicians took the side of the homeowners. New York senator Charles Schumer made a public statement that, in effect, warned NHC not to change its decision on Sandy.[76] New Jersey governor Chris Christie went further, issuing an executive order prohibiting insurance companies from charging hurricane deductibles for Sandy. It wasn't a hurricane, and should stay not a hurricane, by government fiat.[77]

At the same time, NOAA ordered an internal review of the processes that had inhibited NHC from issuing hurricane warnings for Sandy's landfall. This review was politically charged from the outset. The review committee originally included members from the private sector and academia, but it was then reformulated to include only federal employees. Sandy's post-tropical label and that label's consequences were at the center of the review.

The NHC forecasters make their judgments on scientific grounds. How their decision influences insurance deductibles is a societal question, outside their purview. They look only at the meteorological data. But as their entire operation was preparing for a highly publicized review, at least two politicians of national stature, Christie and Schumer, Republican and Democrat, went on record to tell them what the answer should be. Their final judgment, months later, was consistent with their initial one. There is no reason to question it—no established tropical cyclone expert, to my knowledge, has offered any substantive criticism of NHC's classification of Sandy as post-tropical at landfall. But a storm's precise classification had never received this level of scrutiny.

Sandy's political impact went far beyond insurance. National elections, including a presidential one, were just a week away. Sandy was a new kind of October Surprise, this one completely unplanned, and truly surprising.

So much of what modern politicians say is, whether by necessity or habit, scripted and poll-tested. We rarely are convinced we're seeing their true thoughts and feelings. Sandy was sui generis, a giant natural disaster that didn't quite fit any existing script. Those in power had no choice but to react, and to speak about the meaning of what was happening, without having had time to process it. I had the feeling that I could hear the sound of gears turning in their heads.

One of the more remarkable developments was the close and friendly collaboration between Republican governor Christie of New Jersey and the president. Perhaps most surprising were the kind words Christie (not known for tact, warmth, or bipartisanship at this level) had for Obama, repeatedly and publicly. Some viewed those words as the result of political calculation. If so, it didn't look or sound like that. The governor just appeared genuinely appreciative that the U.S. president was working hard to help his state in a time of dire need.

Perhaps most striking of all were the statements of several of our most prominent leaders about the relationship of the storm to climate change.

On Tuesday, October 30, with the floodwaters barely receded from his state, Governor Andrew Cuomo of New York said, "There have been a series of extreme weather events. That is not a political statement; that is a factual statement. Anyone who says there is not a change in weather patterns is denying reality. We have a new reality when it comes to these weather patterns."

He didn't use the words *global warming* or *climate change*, or even the word *climate*, but his meaning was clear enough. Knowing nothing but his party affiliation as a Democrat, one might have guessed that Cuomo would take climate change seriously. But climate change had never been high on the governor's agenda; he had not said much about it in the past. This statement was new and raw.

Mayor Bloomberg, on the other hand, had been a leader on climate change, and on environmental issues more broadly, for years. He had made reducing carbon emissions a priority for the city, and advocated the same internationally, through his leadership in the C40 Cities Climate Leadership Group. He had also taken seriously, to an extent greater than any other local politician in the United States—if not anywhere—the challenge of preparing to adapt the city's infrastructure to the impacts of global warming. He commissioned studies over a period of a decade and established a new city agency, PlaNYC, to address this and other environmental challenges.

Bloomberg's position on climate change, as on many other issues, would have seemed to identify him as an ally of the Democratic Party. But his own party affiliation was Republican (though that means something a little different in New York than in most other states), and his pro-business orientation had led him to take a negative view of Obama, particularly after the president made some remarks, at the height of the financial crisis a few years earlier, that Bloomberg viewed as demonizing the rich. At the same time, Bloomberg's views on many other issues were at odds with the positions of the Republican presidential candidate, Mitt Romney. A week before the election, Bloomberg, a figure of national stature as much as any New York City or State politician had ever been, hadn't endorsed either candidate.

On Thursday, the third day after Sandy made landfall, Bloomberg changed his mind. He announced that he was endorsing Obama for president. He was a single-issue voter. He had problems with both candidates, but he had decided that climate change was overwhelmingly important, and Obama was the candidate more likely to do something about it. Bloomberg wrote,[78] "Our climate is changing. And while the increase in extreme weather we have experienced in New York City and around the world may or may not be the result of it, the risk that it may be—given the devastation it is wreaking—should be enough to compel all elected leaders to take immediate action."

When I heard this, I was stunned. How often does a single event cause a high elected official so quickly, dramatically, and publicly to change his position, particularly on such a high-stakes issue as a presidential election that is a week away? More than that, this was personal for me. The world-famous mayor of my city was publicly changing his mind on the central question before the nation because of my issue (by professional interest as well as political inclination), after an unprecedented disaster right here in our hometown, which we had watched for days until it finally happened, every bit as awfully as we could have imagined from that first day, the week before, when it appeared in the first model runs. After four years of very little talk and even less action on climate change from the federal government, it was more than refreshing, it was wonderful, to see the issue get this attention, and with this result—except that it was actually horrible, because of what had had to happen first.

The following Tuesday, of course, Obama would be reelected by a large margin. It would probably be wrong to give Bloomberg too much credit for this, though his endorsement couldn't have hurt. But it's quite likely that, as many said, both before and after the election, Sandy helped Obama more substantially than that. He was in the news for a positive reason, touring the damage and marshaling all the resources of the federal government in the response and the recovery, instead of exchanging negative remarks with Romney. Romney himself, not holding any office at that time, was nowhere to be seen.

Beyond the positive coverage and visibility for the president himself, Sandy demonstrated better than nearly anything else could have why a strong federal government is important enough to be worth paying taxes for. That message favored Obama. And beyond that, the contrast between the federal response to Sandy under Obama and that to Katrina under George W. Bush was immediately and starkly evident. That favored Obama, too.

There is no doubt that Sandy benefited Obama politically at a critical moment. Most likely he would have won the election anyway; the polls had been consistently in his favor for months. But Sandy's dramatic intrusion into the last stages of the campaign felt transformative. If this storm could threaten to sway a presidential election, perhaps it could also induce broader and more lasting changes to the way we perceive and respond to long-term environmental risks.

PART TWO
WHAT DOES IT MEAN, AND WHAT SHOULD WE DO?

22

THE CLIMATE IS WARMING

The Bottom Line

Let's start from what we know.

The earth's climate is warming due to human emissions of greenhouse gases.

The basic physics of this has been understood for over a century. In the last few decades, as the issue has been made more urgent by the steadily rising temperature, climate science has expanded into a mature field. There has been a great deal of study of many of the complexities.

There is no longer any significant scientific debate about the basic fact of warming. The scientific "consensus" that human-induced global warming is real is not the result of a tallying of opinions. It is the hard-won result of long and careful study by a large community of scientists. Some people—I prefer the term *deniers* to *skeptics*—still claim that there is significant doubt about the basic conclusion that our greenhouse gas emissions are significantly warming the planet. This is simply false.

Claims that the science is in doubt are not based on credible scientific research that contradicts the fact of greenhouse gas warming. There is no such research. These claims are supported by arguments to the effect that the warming isn't occurring, isn't a result of human greenhouse gas emissions, or isn't understood well enough to draw any strong conclusions; but all these arguments are specious, based either on ignorance (sometimes willful) or outright intellectual dishonesty, driven by a dislike of the economic or political implications of human-induced harm to the global environment. These arguments have been refuted, all of them, repeatedly, over many years.

The earth's surface would be entirely frozen without the greenhouse effect. The trapping of infrared radiation by greenhouse gases is the only reason the planet is habitable. There is no doubt whatsoever about this. Greenhouse gases make the climate warmer. So, as we put more greenhouse gases in the atmosphere, the planet gets warmer still. All the possible reasons anyone has thought of that might change this basic answer turn out to be wrong.

There are far too many dimensions to the science of climate change to go through here. I assume that you, reader, accept the science.

Water Vapor

It is worth explaining one key piece of the physics, though. Here's some of what we know about water vapor's role in climate.

Water vapor is crucial for making the warming (both the natural greenhouse effect that keeps the earth habitable and the additional human-induced global warming due to our own emissions of greenhouse gases) as large as it is. Explaining why that is the case—and yet why carbon dioxide and a few other much rarer gases, rather than water vapor, fundamentally drive the warming—will give at least a little flavor of what some of the legitimate scientific arguments around global warming have been about in the past, and how they have been settled. And because water vapor and tropical cyclones are so intimately connected, knowing something about water vapor will help us understand some of the latest research on how a warming climate will affect tropical cyclones.

Water vapor is the most important greenhouse gas. In terms of the total amount of infrared radiation trapped, it does more of the job than CO_2 or any of the others.

But water vapor has a short lifetime in the atmosphere; it is always being put into the air by surface evaporation, and then condensing out again in rainfall. Because of this, water vapor doesn't really control the climate, despite the power of its greenhouse effect. The longer-lived greenhouse gases, of which carbon dioxide is the most important, control the climate. Water vapor acts as a *feedback* that amplifies what carbon dioxide does. CO_2 plays the tune; water vapor is the sound system that cranks up the volume.

Let's imagine we start with an atmosphere that has no CO_2. The climate would be very cold. Then we add CO_2 and see what happens.

First, the CO_2 warms the climate a little.

Then, because the climate is warmer, the amount of water vapor that can be in the air (without condensing and falling out) increases a little. But water vapor is also a greenhouse gas, so the climate warms more.

But then even more water vapor can stay in the air, and so on, until a new, warm equilibrium is reached.[79] That's the "water vapor feedback." It is the main reason global warming is considerably worse than it would be if carbon dioxide (and other relatively long-lived greenhouse gases such as methane) were the only factor. It's also the reason our climate is naturally as balmy as it is in the first place.

CO_2 is the controlling factor because it lasts a long time in the atmosphere. Put it up there, and it stays up there. Water vapor, on the other hand, is always going in and out of the atmosphere. A typical molecule of water stays in the air only a week or two between when it leaves the surface as vapor and returns to it as rain. The amount of water vapor in the air can change quickly; CO_2 can't. So if we add CO_2, the water vapor can respond right away. Add water vapor, on the other hand, and CO_2 doesn't change.

I skipped an important step just now in explaining why there is more water vapor in warmer air. Raise the temperature—say, due to CO_2—and what increases is the *saturation specific humidity*. That's not the amount of water vapor in the air; it's the maximum amount that *can* be in the air. Just because the amount of water vapor in the air *can* increase in a warmer atmosphere doesn't mean it *will* increase. The actual amount of water vapor in the air is what matters for the greenhouse effect.

Outside clouds, the actual amount of water vapor in the atmosphere is always less than the maximum it can be. The relative humidity in clear air (which is most of the sky, most of the time) is less than 100 percent. Otherwise, it wouldn't be clear air; there would be a cloud there. In much of the atmosphere, the relative humidity is not just less, but much less than 100 percent. If the actual amount of water vapor in the air isn't reaching its maximum value, how can it matter what that maximum value is? Isn't it like being in a traffic jam on a highway where the speed limit is irrelevant because no one is going anywhere near it?

Theoretically, it is possible that the amount of water vapor could stay the same or even go down in a warmer climate. If the water vapor were to stay the same under warming, which would imply a decrease in the relative humidity, there would be no water vapor feedback. We would have just the warming due to CO_2 and other long-lasting greenhouse gases, which would be much less than we expect with a realistic water vapor feedback. If it were to decrease, which would imply an even bigger decrease in relative humidity, there would be a *negative* water vapor feedback. The reduction in greenhouse warming by water vapor would offset the warming of carbon dioxide, reducing instead of amplifying the total warming.

There is just no good evidence that this will happen, and lots of evidence that it won't.

To the contrary, multiple lines of evidence indicate that over most of the atmosphere, the relative humidity stays about the same as the climate warms. It goes up a little in some places, down a little in others, but not enough to change the basic conclusion: the temperature regulates the amount of water vapor in the atmosphere in about as simple a way as I described it initially. The warmer atmosphere can hold more water vapor, and does, even though the saturation limit is not reached in most places. As a consequence, the water vapor feedback amplifies climate changes induced by carbon dioxide (or anything else) in the way we expect.

How do we know this?

Climate Models

One important source of information is climate models. Climate models are similar to, and descended directly from, numerical weather prediction models. Climate models do, in fact predict weather. They just predict it for a much longer time than those predictions can be literally accurate, day by day. If a model starts with the real weather on a particular day, due to chaos the model weather will quickly drift away from the real weather. But climate models are not trying to predict the real weather, day by day. They are trying to predict only the average weather over long time periods. While the weather is predictable for only a couple of weeks at most, changes in the average can be predicted much farther ahead.

If you doubt that climate is predictable far ahead even though weather is not, recognize that we all make climate predictions successfully all the time, without thinking about it. We know in the middle of winter that it will be warmer next summer, even though summer is months away. The cause of this change (the earth's orbit bringing our hemisphere around so that it tilts toward the sun instead of away) is what climate scientists call radiative forcing, in the same category as changes in greenhouse gas concentrations. Seasonal changes in the intensity of sunlight occur mostly in the visible part of the electromagnetic spectrum, whereas greenhouse gases act in the infrared; and the switch from winter to summer (in New York, say) involves a much larger change in surface radiation than that caused by human-induced greenhouse gas increases. But the essence of the two is the same. Both are increases in energy input by radiation to the surface that push the climate persistently in one direction: warmer.

Modern climate models differ more deeply from weather models in that they don't simulate just the atmosphere. They also simulate the ocean, the land surface, vegetation, sea ice, and other parts of the earth that influence climate. The modern name for these models, in fact, is "earth system models" rather than "climate models," because the range of processes they simulate has become so much broader than just climate. The laws of physics are used to predict all these processes. The laws can't be used in their pure forms; compromises have to be made because of low resolution—a single pixel, in a typical model's digital image of the atmosphere, measures at least ten and sometimes as much as two hundred miles on a side—or because something other than pure physics as we know it is involved. Vegetation, for example, is represented in modern earth system models, and involves biology and ecology, whose laws are not known so precisely.

Climate modeling is different from weather prediction also in that small differences add up over a long time. If the clouds in a weather prediction model are in the right place but just slightly too thin, it won't matter much. But if they are consistently too thin in a climate model, so that a little too much sunlight gets through, that could produce a significant warm bias (a persistent error in the climate that could really matter). A lot of the work in climate model development is trying to get those small details right.

When CO_2 is added to climate models, the models all produce warming. How much warming varies somewhat, between three and seven or so degrees Centigrade (about double that Fahrenheit) increase in the global mean temperature if CO_2 is doubled. Even the low end of that range is much more than CO_2 alone would produce.

In every serious modern climate model, when the global temperature increases, the amount of water vapor in the air increases with it. That's why the models warm as much as they do. The water vapor increase is about at the rate necessary to keep the relative humidity (the ratio of the actual water vapor content to the maximum possible) constant. In other words, even though most of the atmosphere is not saturated (relative humidity is less than 100 percent), the saturation value, which increases with temperature, does control the actual humidity.

Understanding the Physics—Last Saturation

Why does relative humidity stay nearly constant? When models tell us something so consistently and persistently, it behooves us to understand why. If we can understand the mechanisms that control what the models are doing, we can get insight into whether we should believe them. Are they doing it for reasons that are consistent with what actually happens in the atmosphere? Or because of some persistent flaw that pervades all the models?

There are places in the atmosphere where saturation occurs fairly often. These are the cloudy, rainy places: the tropical monsoons and oceanic rain belts, and the storm tracks of the higher latitudes. These are places where air tends to be rising, either in cumulus cloud updrafts or in the larger-scale ascending zones of extratropical storms. The condensation wrings water vapor out of the air at high altitude, and precipitation brings the water back down to earth in liquid (or solid) form. The air is left dry and cold, high in the atmosphere. In these high, cold places, as long as they remain places where clouds form frequently, we expect that any increase in temperature will lead directly to an increase in humidity, since the humidity is being directly limited by saturation.

Sooner or later, though, the air has to sink to make room for new updrafts. This happens particularly in the subtropical regions, where most deserts are.

As air descends over deserts, the pressure increases and the air warms, but—as long as it doesn't hit a cloud, which is unlikely in a dry zone—no water is added to it. So the water vapor content (the specific humidity) stays the same in a given bit of descending air. The relative humidity drops during the descent. The warmer air could hold more water vapor if it could get it, but none is to be found. The specific humidity was imprinted the last time the air was saturated. It remembers flowing out of the cold top of its last storm, carrying the low humidity that was imprinted on it then, even when it's floating in clear blue skies over a desert.

When the whole climate warms, the temperature in the rainy updrafts increases in the same way it does everywhere else, including in the dry zones. So even when air has descended far into a dry zone, its specific humidity is a little higher than it would have been in the colder climate, because it was imprinted in a warmer storm. The dry zone's temperature has also warmed. The relative humidity (the ratio of the specific humidity to the saturation, or maximum possible humidity) stays about the same, as it works out. There are changes, but they are small. As far as the overall water vapor feedback on climate, constant relative humidity is fairly accurate.

In the 1980s and '90s, there were some strong arguments made that the positive water vapor feedback could be an artifact of bad climate models. At the time, these arguments had merit. Some of the early climate models *assumed* the relative humidity would stay constant, because they weren't sophisticated enough to do anything else; they didn't have realistic clouds or simulate the water cycle very well.

For decades, however, the models have been more sophisticated than that, simulating water vapor and clouds with ever-increasing fidelity. But then, it was still fair to argue that the clouds in the models might not be right. They weren't direct consequences of the laws of physics.

Because almost any real cloud is much smaller than the pixel size in a climate model, they can't be simulated explicitly; instead they are *parameterized*, which means "simulated with some physics but also some imperfect assumptions about how clouds should behave." Parameterized model clouds were (and still are) known to be flawed.

Clouds and water vapor are closely related, since clouds form from vapor; so if the model clouds are bad, one might reasonably expect model water vapor to be bad, too.

Model clouds are much better than they used to be. But more important, we know now that their details don't matter to water vapor nearly as much as one might have thought. In any climate model worth its salt, the air isn't in clouds most of the time—just as the same is true in the real atmosphere. Air in a model pixel containing cloud is pretty much saturated, but then, when it leaves, the clouds don't matter anymore, because the water vapor is imprinted already and the air can't feel a cloud's presence except by being in it. If the model's circulation and temperature structure is anything like that of the real atmosphere, it will be filled mostly with relatively dry air that was imprinted with about the right humidity some time ago.

There are some qualifiers in that paragraph, such as "pretty much," "about," and "as much as one might have thought." But every model makes its clouds a little differently, using slightly different parameterizations, and no one has been able to make a plausible model that doesn't increase water vapor as the climate warms. If the details of clouds were really important, one would be able to find some way to do the model clouds so that water vapor wouldn't increase with warming. No one has found it.

Testing Against Observations

We don't rely only on the models for our understanding of water vapor. We rely on observations as much as we possibly can. The climate has already warmed about one degree Celsius since the early twentieth century. The water vapor in the real atmosphere has kept pace, increasing so as to keep relative humidity roughly constant, confirming our expectations. If it hadn't, in fact, we wouldn't have experienced as much warming as we have. CO_2, despite being the primary driver of the long-term trend, couldn't have done it alone, without the water vapor feedback.

The same thing happens in natural climate variations such as El Niño, which can also warm the planet by as much as a degree (and then cool it down again when the event ends). The wet and dry zones shift around in an El Niño, but on average, globally, the relative humidity doesn't change much. Even detailed tracking of air masses to see where the water vapor comes and goes confirms the picture I described.[80] Humidity is imprinted on the air in clouds and then stays as the air descends, and global temperature increases lead to global humidity increases.

Climate models are marvelously sophisticated, ever-improving things. They produce remarkably vivid and realistic virtual climates. They are direct descendants of weather prediction models, inheritors of the tradition of using physics to understand the atmosphere. We don't trust them fully, but when they tell us something important (especially when they all tell us the same thing), we take them seriously. We do everything we can to compare them to observations and to understand why they are doing what they are doing.

Water vapor is just one of many things that, according to climate models, control the response of climate to greenhouse gas increases. It's a good example because it's important: water vapor is the single most important factor making the warming as strong as it is. We have gone through a long process of questioning whether the models are giving us the right answer about water vapor. We have come to the conclusion, from multiple lines of evidence, that they are.

When we look at other aspects of climate, we have to take the same balanced approach. We ask what we can see in the observations already, what we expect based on our understanding of the physics, and what the models tell us. In some cases, as with water vapor, everything leads to a consistent picture. Then we're pretty certain we know the answer. In others, we are left with contradictions, or gaps in our understanding.

When we ask how Sandy was related to climate change, the answer has many different parts. They range from "quite certain" to "not a clue."

23

THE SEA IS RISING

As the planet warms, the sea rises.

Global sea level has been increasing slowly for the better part of the last century. In the last couple of decades, rate of rise has been about three millimeters per year. That is not a big difference from one year to the next, but it is steady. Three millimeters every year will add up to about a foot in a century. That's not nothing, but the real risk is that the rate of rise will accelerate dramatically.

Why does it rise?

Much of the rise so far is a result of thermal expansion. Nearly all matter swells up when its temperature rises; saltwater is no exception. As the oceans warm, the same amount of water takes up a little more space. The expansion is very slight, but because the oceans are deep, it adds up.

Thermal expansion is a relatively predictable process. We know very precisely how much any little bit of water will expand for any particular increase in its temperature. There is still some uncertainty as to the rate of expansion of the whole ocean. In part this is because we don't know exactly how fast the warming, which starts at the surface, will be brought down to the deeper ocean by mixing of deep and near-surface waters. And of course we don't know exactly how quickly the warming itself will occur. That depends not only on uncertainties in climate physics (as expressed, for example, in the difference in the warming produced by different models for the same greenhouse gas increases), but also on the greenhouse gas emissions themselves, which depend on economics and politics.

Still, though, we can place reasonable bounds on how quickly thermal expansion can proceed. By itself, it won't be catastrophic, unless the rate of warming itself is much faster than we expect, in which case we will have many other huge problems.

The truly frightening unknown in sea level rise, though, is how quickly the ice on land will melt.

Sea ice melting, we should first clarify, is irrelevant to sea level rise. Sea ice is already, by definition, floating. Just as ice cubes in a glass of water don't raise the water level in the glass when they melt—try it—melting sea ice doesn't raise the water level in the oceans. Though a well-publicized conse-quence of global warming, and one with many other serious consequences, sea ice melting doesn't affect sea level.

When ice on land melts, on the other hand, water ends up in the sea that wasn't there before, so the sea level does rise—that's pouring more water into the glass.

By far most of the ice on land is in Greenland and Antarctica. Glaciers elsewhere add up to very little compared to those two. The problem of pre-dicting the rate of sea level rise is mainly the problem of predicting how rapidly the Greenland and Antarctica ice sheets will melt. That is a problem for the glaciologists, and it is difficult.

If it were simply a matter of taking a small chunk of ice, asking how rapidly it will melt for a given increase in temperature, and then multiply-ing the answer by a huge number to scale it up to the size of a continental ice sheet, that wouldn't be too bad. Just as we know how quickly seawater expands if we warm it, we also know how rapidly ice melts if we warm it, just from thermodynamics alone. We would still have to predict the rate of warming, but we have to do that with all predictions of any downstream impact of climate change.[81] The problem is that the behavior of a large ice sheet can't be predicted easily just from the physical properties of small pieces of ice.

An ice sheet is a dynamic thing; it flows, slowly, under gravity. Snow on top makes it bigger. Over time, the snow is compacted down to solid ice by the weight of the snow above it. Meanwhile, the ice at the margins edges out into the sea, to be replaced by the slow downhill flow of the ice above and inland. If the loss of ice to the sea is more rapid than the accumulation of snow at the top, the ice sheet shrinks, and the sea level rises.

The flow of the ice is slow but still difficult to predict. It is governed by not just the static properties of the ice, but also its dynamic properties: how it moves. And not just the ice itself, but what happens at the bottom, between the ice and the rock or soil below, matters. The high pressure from the ice's weight can cause a little melting at the bottom, despite the extreme cold. Any liquid water can lubricate the ice's flow and accelerate it. This process, like others going on in the ice sheets, is very hard to observe—the bottom of an ice sheet is an inhospitable place to make measurements—and the conditions cannot easily be reproduced in the laboratory. As in all of climate science, computer models are necessary, but the models of ice sheets are in a state of relative infancy compared to the models of the atmosphere and ocean.

When the large international group of scientists writing the "Fourth Assessment Report (AR4) of the Intergovernmental Panel on Climate Change" made its estimates of sea level rise in 2007, they knew they didn't know enough about the dynamics of ice sheets to make a confident prediction about how much larger the ice sheets' contribution to sea level rise would become in the next hundred years.

There was evidence that some glaciers might be dramatically accelerating, surging forward after icebergs previously pinned at the front, holding back the ice behind, had calved into the sea. But the observations of this were new. It wasn't clear yet just how fast the glaciers could move, or whether the fastest rates of motion would be short-lived or could be sustained for long periods of time.

Not knowing how to deal with it, the Intergovernmental Panel on Climate Change neglected it—in AR4, the IPCC assumed that sea level rise would come just from thermal expansion, and from ice sheet melting at the rate it had been for the decade between 1993 and 2003. It came up with projections of sea level rise for the next century between twenty and fifty centimeters, or from less than a foot to, at the most, a foot and a half.

It is almost certain, though, that the ice sheets will melt more rapidly in the future than they have in the past. In the last decade, the flow of ice has been accelerating in Greenland—both at the margins, where the ice flows out into the sea, and in the interior. At the edges of both Greenland and Antarctica, previously unexpectedly enormous chunks of ice have broken off, in events violent enough to cause moderate earthquakes that could be detected globally. (My colleague Meredith Nettles, at Columbia's Lamont-Doherty Earth Observatory, has been studying these for around a decade.) We still don't know enough, specifically about how ice sheets work, to make accurate estimates of how rapidly these processes will accelerate, but we can't safely assume that they won't accelerate at all.

Current estimates of sea level rise for the next century are still highly uncertain, but the range in the new, Fifth Assessment report of the IPCC is between forty-eight and ninety-six centimeters (basically, between a foot and a half and three feet). Many experts believe these estimates are still low; certainly the trend over time is for the estimates to increase. The primary scientific literature on the topic suggests that estimates as high as two meters (six feet) by 2100 have to be taken seriously as possibilities.[82]

It would be enormously difficult for human society to adjust to six feet of sea level rise in a century. That would place many heavily populated coastal regions underwater. But even that rapid rate of rise would still be just a down payment. It takes a long time for the heat at the ocean's surface to mix down to the deep ocean, and a long time to melt ice sheets—much longer than it takes to warm the average surface climate of the earth. Sea level rise will lag behind warming, so what we see in the first hundred years will be just the beginning, even if the warming stops after a hundred years.

One recent estimate is that each degree Celsius (a little less than two degrees Fahrenheit) will, in the end, lead to 2.3 meters (more than 7 feet) of sea level rise.[83] Since we expect something on the order of three to eight degrees Celsius of warming over a century, that's 20 to 60 feet of sea level rise.

These numbers sound crazy; but we know they are not. Sea level has risen and fallen by that much before. In the last ice age, when glaciers covered much of North America, sea level was low enough for ancient humans to walk from Asia across what is now the Bering Strait to North America. A large research project by my colleague Maureen Raymo at Lamont, using a wide range of geological evidence, is reconstructing sea levels in the distant past. She and her colleagues have shown that in a warm period around 125,000 years ago (when the concentration of carbon dioxide in the atmosphere was about what it is today, and the earth's orbit was such that the sunlight on the earth's high latitudes was slightly stronger than today, melting the ice caps), sea levels were as much as nine meters (almost thirty feet) higher than today.[84] Still farther back in time, around 400,000 years ago, global average sea level may have been as high as ten to twelve meters higher than today.[85]

While such enormous increases in sea level are truly horrifying to consider, they won't be achieved anytime soon. It will probably take at least a few centuries, maybe as long as two thousand years. But recognizing them as very long-term possibilities makes it much easier to grasp that six feet in a century is not far-fetched at all.

What we know for sure is that sea level is going up, not down. There is no realistic scenario under which the ice sheets will grow as the planet warms; they can only shrink, and there is only one place for the meltwater to go. Thermal expansion, too, goes only one way. When we ask how Sandy is related to climate change, this is the easiest part of the answer.

Everything I have said so far about sea level rise is really about its global average. But sea level rise is not uniformly distributed around the earth. The sea will rise much more in some places than in others. Some of the differences are due to differences in ocean warming from one place to another. Some are due to changing winds, which change the ocean currents, which in turn distort the height of the sea surface. (In the ocean, as in the atmosphere, pressure and flow are related nearly geostrophically, so that currents are associated with high- and low-pressure systems, and these are accompanied by high and low elevations of the sea surface.) As the ice sheets melt, surprisingly large distortions of the sea surface will occur due to changes in the earth's gravitational field. A giant ice sheet exerts a gravitational pull that draws the sea toward it. When the ice sheet is gone, the sea will slosh away, and the sea level near the former ice sheet can drop quite a lot, relative to the global average.[86]

In New York City, sea level has risen around a foot in the last hundred years. Most of that is due to warming.

A few inches of it is not really true sea level rise at all, but apparent sea level rise due to land sinking. This is natural, part of our ongoing recovery from the last ice age ten thousand years ago. The Laurentide Ice Sheet covered virtually all of what is now Canada and the northern tier of the United States. Modern-day New York City was at the edge of it; Long Island is the terminal moraine, the pileup of rocks and junk pushed out by the glacier's leading edge. At the center of the ice sheet, the massive weight of two miles of ice pushed the land surface down under it, creating a depression. New York, at the edge, was pushed up by the flexing of the earth's crust. Then the ice melted. The land has been rebounding: rising in Canada where the ice was thickest; sinking here in New York City.

The total water level in Sandy was the mean sea level, plus the tide, plus the surge. Had the storm occurred a hundred years ago, the water would have been about a foot lower, with most of that difference due to warming. It would have been equivalent to lowering the surge from nine to eight feet, the total storm tide from fourteen to thirteen feet above the low-tide mark.

That difference is enough to matter. Irene, at nine feet above low tide, didn't flood much of anything in the city. Sandy's surge was five feet higher, and it was a catastrophe. A foot is one-fifth that difference.

As we look ahead to greater sea level increases (predominantly due to warming, even more so than in the recent past), the danger is not just that higher seas will make bad flood events worse. They will do that; if the sea had been six feet higher to start with (within the range of possibility for 2100), the total water level produced by Sandy would have been twenty feet above today's low-tide mark, instead of fourteen. It's hard to imagine what that would have done. But worse than that, floods such as Sandy's, already quite bad enough, will become much, much more frequent.

With just three feet of sea level rise, it would take only a six-foot surge at high tide to do what Sandy did. It doesn't take a hurricane to produce six feet. A strong winter nor'easter can do it. In New York and other mid- or high-latitude coastal cities, nor'easters are much more common than hurricanes. Sandy, perhaps a several-hundred-year event for New York City,[87] could become a thirty-year event.

The one good thing about sea level rise is that it's slow. Although several feet in a century would be incredibly rapid compared to what has happened naturally in the past, it's still very slow compared to the approach of a hurricane. We can respond to sea level rise strategically.

For some places, of course, there may be no solution. Low-lying small island states such as the Maldives or the Marshall Islands may simply be wiped off the map.

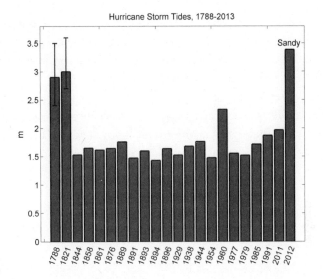

Historical high water levels (meters relative to the mean tide) due to tropical cyclones in New York Harbor. The events are ordered in time although not evenly spaced. Uncertainty estimates are given by the error bars for the first two events; the data become more accurate after 1844. Data after 1920 are from the tide gauge at the Battery, in Lower Manhattan. Sandy is the last bar, in 2012. *(Image courtesy of Stefan Talke and Philip Orton. Data are from S. A. Talke, P. Orton, and A. Jay, "Increasing Storm Tides in New York Harbor, 1844–2013,"* Geophysical Research Letters, *doi: 10.1002/2014GL059574 [2014])*

For coastal zones in the United States and other continental nations, though, there are possibilities. We can build seawalls; elevate things; make buildings and infrastructure more resistant to flooding and more resilient in their recovery from it; or simply move back from the coast, if we have to. And of course if we really wanted to get at the long-term problem, we could reduce our emissions of greenhouse gases.

But any of these solutions requires foresight, will, and resources. So far, we as a species have not been so good at marshaling these things to address long-term environmental risks.

24

ZEELAND

Delta Works

Zeeland is the Netherlands' lowest-lying province, in the country's southwest. It is a giant delta. Three large rivers, the Rhine, Meuse, and Scheldt, run into the ocean in Zeeland; much of Europe's rainwater, and even snowmelt from the Alps far to the south and east, drains there. Originally, high natural dunes lining the coast were the only barriers keeping the sea from the land behind them, much of it below the high-tide mark. This original landscape was transient; the land flooded regularly, and the coastline was perpetually evolving, as the sea and the rivers deposited sediment and took it away.

Over the last millennium, the Dutch increasingly took control of the landscape. The country was covered with dikes, and pumps were put in place to keep the land dry, to protect it from the water seeping in from the sides. New land was taken from the sea. The Netherlands, and Zeeland in particular, is a substantially larger place, in square kilometers, than it was a thousand years ago. But many of those square kilometers are below sea level. They were taken from the sea; keeping them from it is a job that will never be finished.

The dikes are still there; all flood-control structures in the Netherlands, if laid end to end, would stretch for seventeen thousand five hundred kilometers, over ten thousand miles. Many of those kilometers are in Zeeland. Most are the traditional, iconic dikes (linear mounds of clay and peat with grass growing on them) and other similarly old-fashioned, and in some cases genuinely old, constructions. But a small fraction of those kilometers constitutes a different type of thing altogether. The last fifty years have seen a massive technological upgrade to Zeeland's defenses.

A series of floods in the late nineteenth and early twentieth centuries, as the Netherlands was transforming into a modern, industrial nation, served notice that Zeeland's dikes (traditionally maintained by small, local boards made up of those living under their immediate protection, rather than any centralized authority) were inadequate. The federal government commissioned a set of scientific and engineering studies and reports over decades, spanning the period before and after World War II. But nothing major was constructed until the big one hit, on the night of Saturday, January 31, 1953.

The storm was a powerful extratropical cyclone, moving first, while still over the open waters of the North Atlantic, on a track toward the northwest. After the center passed north of the British Isles, it took a sharp right turn toward the southwest. This track brought it farther south, and more directly into the Zeeland coast, than any previously recorded major storm. As the center approached, winds initially from the southwest, in the storm's warm sector, turned sharply to blow from a more northwesterly direction as the cold front passed. The southwest–northeast-oriented coastline then lay perpendicular to the winds, so that the storm surge was driven directly onshore.

The surge grew over the course of the day on the thirty-first. It was a spring tide, with the highest tides of the month. The surge, simulated by subsequent researchers to reach over three meters in many parts of the delta, built and then ebbed slowly enough that it would inevitably be close to its peak at high tide, leading to total water levels of four meters or more above the average sea level.

To a New York meteorologist, post-Sandy, the 1953 flood sounds eerily familiar. The storm wasn't more powerful than others in the recorded past, but its winds blew right onshore, as its predecessors' hadn't, due to its track, which contained an unusual sharp turn. It coincided with the spring tide. These coincidences led to a storm tide higher than had been seen in generations, and an unprecedented disaster. But the natural event itself, the storm and the water level, must have been understood to be in the realm of the possible. Similar floods had occurred in past centuries, as in New York in 1821.

But as with New York in 2012, Zeeland in 1953 was occupied by a civilization radically different from any that might have been in place the last time a flood of that magnitude had occurred, centuries earlier. The modern, dense population had no memory, personal or historical, of a disaster of this magnitude.[88]

The country's storm tide warning service, a section of the Royal Netherlands Meteorological Service that had been created in 1921 just for such events, had known of the storm for a day or so and had begun sending out warning messages by telegram and radio that morning as the storm approached. The seriousness of the messages increased by evening. Many people didn't get them; thirty different local authorities had subscribed to the warning messages, but many in the storm's path had not. Worse, it was the weekend, and the peak waters didn't come until the first hours of the morning on Sunday, February 1. People were sleeping, the radio didn't broadcast at that hour, and there weren't many telephones. Zeelanders awoke to the sea pouring through broken dikes.

One thousand eight hundred and thirty-five people died in the 1953 flood. Something like a hundred thousand were evacuated. The head of cattle killed numbered at least in the tens of thousands, possibly hundreds of thousands. The relief efforts were chaotic at first, but became more organized as days passed. Volunteers came in by train; the Dutch military, then aided by the French and American ones, used amphibious vehicles, helicopters, and boats to get people out and food into remote island villages.

Reconstruction of the destroyed areas was a longer-term project. With many dikes breached, water would not leave the areas below sea level on its own. The sea flowed in and out with the tides. Repairing some of the larger breaches required major engineering efforts. The last breach was closed nine months after the storm.

Never again. With the technical ground prepared by the earlier generations of studies, and the impetus that only the hard experience of a recent major disaster can give, the government decided to do something serious. The Delta Committee was formed on February 18, with the flood less than three weeks in the past. Thirteen of the Netherlands' best scientists and engineers were charged with developing a plan for a set of "Delta Works," hard engineering structures to protect Zeeland from future floods.

Instead of recommending that the dikes be rebuilt higher and stronger, the committee went in a new direction. It proposed closing the large gaps in

Zeeland's coastline entirely. The existing natural barriers, the dunes lining the beaches on the ocean sides of the islands, would be connected by a set of artificial barriers between one island and the next, keeping storm surges out of the inlets.

The Delta Plan was submitted to Parliament in late 1955 and approved for construction in the summer of 1956, around three and a half years after the Big Flood. By today's standards, this was remarkably rapid action. In fact, one barrier was already under construction by January 1954, less than a year after the flood, farther inland, where the Hollandsche Ijssel, a tributary to the Meuse, enters the larger river, just east of Rotterdam. A major surge up the Hollandsche Ijssel would threaten more central parts of the country. (Immediately upstream of the entrance to the Meuse is the town of Gouda, famous for the cheese of the same name.)

Construction of the barriers proceeded over the succeeding decades. By the mid-1970s, several were done, including those on two of the three largest inlets to Zeeland, the Haringvliet and Brouwershavenschegat. These are essentially dams. They include sluice gates that can be opened, either to alleviate flooding on the rivers behind them or to let saltwater in to prevent the rivers from freezing. Apart from those rare circumstances, though, the gates stay closed. Storm surges are kept out, but so are the regular tides. The barriers severed the connections between the inlets and the sea.

The inlets, historically tidal saltwater bodies, became fresh and flat. Their ecologies changed dramatically. The local cuisine is famous for mussels; these couldn't live in the dammed inlets. Nor could oysters, or the saltwater fish that had been there.

The environmental movement was coming into maturity in the early 1970s. Rachel Carson's *Silent Spring* had been published in 1962, and the pesticide DDT was banned in the United States in 1972. (Joni Mitchell sang, "Farmer, farmer, put away your DDT," in 1970.) The environmental cost of closing the Zeeland coast began to weigh heavier than it had twenty years earlier.

In 1975 the last and biggest of the planned Delta Works was under construction, across the Oosterschelde (Eastern Scheldt), in the south of Zeeland. It was being built as a dam like the others. Under environmentalist pressure, Parliament agreed to spend more money than originally planned to finish the barrier in a way that would keep the Oosterschelde tidal and salty, leaving the mussels in place while still stopping storm surges.

Map showing Zeeland province, the Netherlands, and its surrounding areas, including the storm surge barriers built after the 1953 flood, indicated by light gray lines. *(Courtesy of Andrew Kruczkiewicz and Jerrod Lessel)*

The result, finished in 1986 after a huge engineering effort, is a four-kilometer-long series of steel gates, with road bridges on top and an island in the middle. The island, Neeltje Jans, is artificial, and was built at the site where the construction took place; it now has a tourist center with a museum, with an aquarium, a playground, and a restaurant featuring mussels.

Most of the time, the gates stay above the water, allowing the tides to pass underneath. When a storm threatens to push a high surge in from the sea—a forecast of three meters above normal sea level is the trigger point—the gates are closed. The three-meter level is, historically, the six-month flood, meaning it occurs an average of twice per year; the gates have been closed twenty-four times since construction.

With the construction of the Oosterschelde Barrier, the large gaps in Zeeland's coastline were closed. Rotterdam, though, remained vulnerable. The Netherlands' second city, Rotterdam straddles the Meuse about fifteen kilometers upstream from where the river runs into the North Sea at Hoek van Holland ("Hook of Holland") through a series of ship canals. This is the northern boundary of Zeeland. Rotterdam is in the province of South Holland, which begins on the Meuse's north bank.

Compared to the enormous inlets of the Haringvliet, Brouwershaven-schegat, and Oosterschelde, the canals running past Hoek van Holland are very narrow. Storm surges don't get as high as easily—the three-meter water level that is the six-month flood at Oosterschelde is the ten-year flood here. The 1953 flood hadn't been a catastrophe for Rotterdam.

Maybe the lower risk was a reason Rotterdam wasn't covered by the original Delta Plan. Also, though, the country couldn't afford to stop the ships from coming through. Rotterdam was, and is, the largest port in Europe. The other barriers had locks to let ships pass, but these couldn't have handled Rotterdam's traffic.

In the first half of the 1980s, as the Oosterschelde was being built to finish the Delta Plan for Zeeland, a parallel plan, the Europoortkering, was in the works to protect Rotterdam and the province of South Holland. To avoid blocking access to the ports, the first draft of the plan involved dikes: more, bigger, better dikes. The scale of the project would be massive, and would take multiple decades; some of the planned dikes would require relocating ancient towns. With the Oosterschelde successfully done and confidence in creative engineering solutions high, another one was sought for Rotterdam.

The result was the Maeslant Barrier: two hollow steel arcs that normally rest ashore, one on each side, leaving the channel entirely unobstructed. Each arc is connected to a ball joint ("like your shoulder," as the guide who gives the English-language tour puts it) by an enormous wedge-shaped arm made up of two pairs of enormous steel beams that fan out from the joint and are crisscrossed by a web of smaller beams. The entire structure is painted white so that the sun won't heat it up and cause it to expand; in 40-degree Celsius heat (104 degrees Fahrenheit, which doesn't happen much in the Netherlands) the arms lengthen by just twenty centimeters. This is one of many tight and carefully planned tolerances; the ball joint (the biggest in the world, at ten meters in diameter) sitting in a massive foundation sunk into the earth, is designed to give two millimeters when the arcs are closed and absorbing the force of the water.

The barrier closes when a big storm surge threatens. Again, a prediction of three meters above sea level is the number. That's the ten-year flood at Hoek van Holland; since 1997, when the Maeslant Barrier was finished, it hasn't closed, except for tests.

In a real event, the decision to close or not is made by a computer (disconnected from the Internet to avoid hackers) that takes in all the relevant information: observations of water levels, weather forecasts, and such. The decision process is meant to be independent of humans, to avoid the possibility of any political pressures.

The Maeslant storm surge barrier in Hoek van Holland (Hook of Holland), the Netherlands. This image shows the barrier in the closed position, though the arcs are still floating at the surface; to block a storm surge, the two arcs comprising the barrier would fill with water and sink to the bottom to close off the channel. When no storm threatens, the channel is open, and the two arcs rest on their respective banks. *(Courtesy of Rijkswaterstaat, of the Dutch Ministry of Infrastructure and the Environment)*

The arcs themselves are hollow, and float as they are swung out into the channel. Once they are in position, holes open up, allowing them to fill with water so that they can sink to the bottom, shutting out the sea and leaving their tops high enough to stop a five-meter storm tide. Five meters is believed to be the ten-thousand-year flood, the standard used for the Zeeland project.

Zeeland Barriers up Close

During the summer of 2013, while I was writing this book, Marit and I had the opportunity to visit several of the Zeeland barriers. We started with the Maeslant Barrier. The most glamorous in online photos, it is also the easiest of the barriers to reach by public transport.

Up-close views of the barrier don't convey the same drama, at first, as the online photos. Most of those are aerials. From above, one takes in the whole structure at once and sees visually how it works. Even without knowing the details, one immediately sees the creativity of the engineering solution. Standing next to it, one doesn't have that perspective. Without any prior knowledge, it might be difficult, at first, even to figure out what the structure is. From the train, passing right by it, one sees the barrier as blending in with the landscape of factories and port facilities.

What does sink in from up close is the size of the thing. In aerial photos, the straightness of both the roads and the channel itself, the flatness of the scene, and the monotone colors (green lawns and greenish-blue water setting off the white barrier) give the barrier the look of a toy-size engineer's model. Standing next to the arms prohibits that impression. One has to hold the aerial view in one's mind in order to keep the components and their functions in perspective; but their raw dimensions (length, girth, mass) impose themselves.

After the tour, we were fortunate to hitch a ride with two other tourists, Darrel Kerr and Wouter-Jan Lippmann, out on a driving tour of not just the Maeslant but also the other Zeeland barriers.

Driving south through Zeeland, we crossed the Haringvliet Dam, 4.5 kilometers long, from Voorne to Goeree, and then the Brouwershavens-chegat, stretching 6.5 kilometers from Goeree to Schouwen. Both these are true dams, turning the inlets behind them into lakes, to our east, dotted by a few sailboats. We passed beaches on the sea side, to the west.

We came last to the Oosterschelde, the grand finale of the original Delta Plan, spanning the nine kilometers between Schouwen and Noord Beveland. We drove over the first section, pairs of narrow towers rising out of each concrete pier in the long series lining the right side of the bridge, to land on Neeltje Jans, the artificial island in the center of the barrier, where the structure was built on site.

The engineering of the Oosterschelde Barrier has layers upon layers, both literally and figuratively.

The seabed, too soft naturally to handle the forces the structure would place on it, was first compacted to make it firmer: giant steel "needles" were sunk from a custom-built ship down into the sand and vibrated, causing the sand to settle and squeeze out water from the tiny gaps between the grains, down to a depth of fifteen meters below the sea's bottom. This took three years.

A broad trench was then dug in the compacted sand, and two different layers of "mattresses" (polypropylene filled with gravel) were laid down. The mattresses were built in a custom factory on Neeltje Jans. One custom-built ship carried them into position, and another ship (different, also custom-built) assisted in lowering them and sealing them into place. All this was done during one-hour slack-tide periods, when currents were weak. Tight tolerances were monitored by underwater cameras fed to a control center full of video screens and 1980s-era computers.

Each concrete pier was constructed in a pit, with a bottom fifteen meters below sea level, continuously pumped to keep it dry, on the new (also custom-built) artificial island of Neeltje Jans. Yet another custom ship was designed and built to move the piers from the pit to their final locations, dropping them in (again, at slack tide) and positioning them precisely. The hollow concrete piers were filled with sand to weigh them down, cemented to the mattresses, weighed down at their bases by thirty-ton bags of stone and asphalt, and the whole base embedded in a sill of stones, the largest weighing ten tons, the top layer put into place by still another custom-built ship.

With the concrete piers in place, each one forty-five meters from the next, the steel gates were then slotted in, along with all the supporting steel-and-concrete structures. At the end of the barrier adjacent to Neeltje Jans, tourists can climb in and explore the first two piers.

Even more than with the Maeslant, proximity to the Oosterschelde Barrier doesn't begin to do justice to its size. One can't come close to seeing the full extent of it: nine kilometers, over five miles. Because it is built in many short sections (each with its black steel gate, held between its two gray concrete piers, each in its enclosing cage of access ducts, balconies, and stairways), one is drawn, when close up, to the detail of each of its individual substructures. I was transfixed by the tide, flowing fast just below the black steel gate.

The Will and the Way

The Dutch are the world's experts in flood management. They have lived below sea level for many centuries, and have thought more about flood prevention than anyone in the developed world. They have learned how to deal with it, by a combination of practicality and hard experience.

This expertise is now a national export. Go anywhere flood-prone, and you will find Dutch experts. Some of them were studying New York City before Sandy; the risk was apparent enough. After Hurricane Katrina devastated New Orleans in 2005, the Dutch solutions suddenly were taken much more seriously in the United States.

Post Katrina, the U.S. Army Corps of Engineers constructed a new set of structures to protect New Orleans against future hurricanes. In April 2014, I had an opportunity to see the largest of them, the Inner Harbor Navigation Canal Lake Borgne Surge Barrier.[89] Finished in 2013, it's a nearly two-mile-long wall rising nearly thirty feet above the water. A single gate near the midpoint stays open in calm weather, allowing boats through. (Because only this one gate was included—rather than many, as in the Oosterschelde—the waters on the inland side are mostly cut off from the tide, and have become much fresher, putting an end to formerly good fishing and shrimping on that side.) It took Katrina to get it built, even though New Orleans was well known, for many years, to be exceptionally vulnerable to flooding by hurricane-induced storm surges. The previous hurricane system—the one that failed in Katrina—had been built by the Army Corps after Hurricane Betsy in 1965 caused serious flooding (by pre-Katrina standards) and over a billion dollars in damage.

Outside New Orleans, the Dutch solutions faded from our American view again in the succeeding years, only to return in late 2012. Many post-Sandy disaster-inspired stories about Dutch barriers seem to imply that if we are serious about wanting to do something ourselves in New York City, we will follow the Dutch example, and think big.[90]

Much of our time's highest-profile engineering is virtual, in software. The Dutch barriers are old-school, hard-core engineering in its primitive essence, despite their obvious modernity. They keep harsh nature out of the tenuous shelters of people. They require, quite literally, the heaviest lifting we can manage, to meet the water at its own scale.

Making Room

Large structures inspire awe, but modern Dutch flood control has developed other dimensions. Newer projects in particular have treated the sea less confrontationally.

The Zandmotor (sand motor) is a massive beach-replenishment project. Sand from far offshore is sprayed onto shore by boats; the sand is deposited in one place, where it forms a hook-shaped bar extruding from the beach. The shape and location are designed so that wind and tides will distribute the sand along the beach over subsequent decades.[91]

The "Room for the River" plan is, in some respects, an even more radical project than the giant barriers were, though the ambition that makes it so is political rather than technical. The plan involves removing inhabitants, mainly farmers, from land in the immediate floodplain. In a flood, driven either by storm surge or rainfall, the river can simply inundate the vacated land, as it would have done prehistorically, with no harm done.

This strategy is known here in the United States as "retreat." In the Netherlands, when one group of farmers living in an area slated to be returned to the river were told by the government that they had to leave, some chose to oppose the order, and negotiated a compromise that allowed them to stay with their houses, on the few small hills within the area. But most left.[92]

Now it is starting to seem possible that something similar may happen in New York City, on a small scale. In early 2013, a few months after Sandy, Governor Cuomo announced that New York State would offer to buy out homeowners in some of the most badly flooded areas, offering them pre-Sandy value for their homes. If the state can get possession of large blocks of property, that land will be permanently set aside to become natural area, never again developed. If it needs to flood, it can flood. Cuomo said, in his State of the State address in January,[93] "There are some parcels that Mother Nature owns. She may only visit once every few years, but she owns the parcel and when she comes to visit, she visits."

This is not quite the Dutch model. In the Netherlands, farmers were *told* by the government that they had to leave. They banded together, negotiated with the government, and some of them were allowed to stay, on hills. In New York, the state just made homeowners an offer. They were under no explicit compulsion to take it.

In practice, though, the New York model has the potential to become pretty Dutch, if it gains momentum. In Oakwood Beach, Staten Island (a neighborhood not only devastated by Sandy, but prone to rain-driven flooding, and brushfires in the surrounding *Phragmites*-filled wetlands), most residents are taking the deal, leaving those remaining feeling abandoned and pressured as their community disappears.[94] In fact, the program makes a lot of sense only if everyone goes. Mother Nature's visits won't be friendly if any human tenants are still there when they happen.

As novel as it is, and as remarkable as it would have seemed pre-Sandy, the New York buyout plan (and its New Jersey counterpart, which has been allocated three hundred million dollars) is still tiny compared to the scale of the problem. In the overwhelming majority of the places flooded by Sandy, no buyouts have been suggested. No one is seriously proposing making room for the harbor in Lower Manhattan.

So should we perhaps consider the more confrontational Dutch strategy, in which heavy engineering is used to keep Mother Nature out of the room?

Barriers in New York?

In 2009, I received an e-mail announcement for a conference organized by the American Society of Civil Engineers, at Brooklyn Polytechnic University (now part of New York University). The topic was "Storm Surge Barriers for New York Harbor." This was four years after Katrina had raised awareness of flood hazards to coastal cities, and it was clear that there was a need for a more serious discussion of the risk to New York City. I took the subway to Brooklyn.

The presentations were given by engineers and a couple of meteorologists and oceanographers. The first few presentations set the context. Malcolm Bowman, of SUNY Stony Brook, described in detail the complex physical oceanography of New York Harbor, where the Atlantic meets both the Hudson and Long Island Sound, each with different tides and currents. There were talks about lessons learned from Katrina.

As advertised, the focus was on the idea of Dutch-style barriers across New York Harbor. Collectively, the speakers made the case for a giant engineering project to close off the major entrances: the Verrazano-Narrows, Arthur Kill, perhaps even the more than five-mile span from Sandy Hook to the Rockaways.

A major hurricane had not hit New York City in its modern history, but clearly one could. The 1938 hurricane had not missed the city by much. If one like 1938 were to hit, the city would be in trouble. Many low-lying areas were vulnerable to storm surge: Lower Manhattan, the Rockaways, Red Hook, other sections of Brooklyn's and Queens's coasts, and Staten Island. The Dutch barriers, and their counterpart on the Thames, had proved that something big could be built successfully.

To me, as a scientist, this all sounded plausible enough. But to me as a native and long-time resident of the city, it had the air of the fantastic.

One could, of course, debate the costs and benefits. The speakers understood that, taking the reality of the risk as given, reasonable minds might still disagree about whether smaller-scale or "softer" responses might be better than large barriers. But it was clear that the risk was real and that barriers were a potential solution worth taking seriously. Barriers would be technically feasible; this was not a meeting about putting colonies on Mars. What seemed disconnected from reality was the belief, not stated in words but implicit in the meeting itself, that there was any chance the barriers might ever get built.

A *New York Times* article, published a few days post-Sandy—I don't know of any written at the time of the meeting itself—by a reporter who was there, described a small group of engineers eating a dinner of cold spaghetti during a melancholy evening of presentations, entirely out of the media spotlight.[95] I hadn't remembered the spaghetti until I read that, but that detail captured the feeling I remember. As the reporter, McKenzie Funk, wrote, the conference "had the sad air of what was then an entirely lost cause."

It was a group of engineers getting together to talk about something they believed should be built, and had no doubt could be built, given the necessary charge. The charge was entirely absent. No politician of any significance was at the meeting; there were presentations from the staff of Mayor Bloomberg's new PlaNYC office, but these were just about the city's other activities on hurricane risk; they didn't indicate any endorsement of barriers by the mayor.

In February 2013, the American Society of Civil Engineers held another meeting, in the same place. The topic was, nominally, a little different: the damage inflicted by Sandy on the New York metropolitan area's infrastructure. There were detailed presentations by officials from Port Authority, the MTA, Con Edison—all the entities that keep the tristate area physically functioning. Many of the presentations were about describing the damage to the facilities, and what actions were taken before, during, and after the storm. The Con Ed presentation was particularly compelling. Mayor Bloomberg wasn't there, but Mayor Dawn Zimmer, of Hoboken, was. She described in plain language the extreme suffering of her small city and its struggle to recover after the storm.

Storm surge barriers were the real agenda. It was essentially the same group of people, in the same room as in 2009, but the feeling was entirely different. There was a clear sense of vindication. The 2009 meeting now looked prescient. Had the barriers been in place, the catastrophe would not have occurred—at least not for Manhattan.

Besides all the government officials, the meeting was populated by big private engineering firms. The engineers were not just sitting in the auditorium, but standing outside in the hallway, where impressive scale models of barrier technologies were on display. Even the smallest solutions being discussed at the conference would have a price tag in the tens of billions.

There was no joy, per se. Like us meteorologists and other scientists who study disasters, the engineers who try to prevent them understand and respect the tension between excitement about doing their jobs and respect for those who suffer when disaster strikes. Nonetheless, there was electricity in the room. Perhaps something might actually get built now.

25

HURRICANES AND CLIMATE

As the sun shines down on the tropical oceans, they ventilate and cool themselves by evaporating water into the atmosphere, the sun's warmth stored as latent heat in the water vapor. When that water condenses and rains out in tall cumulus clouds, the warmth is transferred back to the atmosphere, warming it up to high levels. The atmosphere cools itself by emitting infrared radiation from those high levels, above the greenhouse trapping of the water vapor below, to space. That is the most basic cycle of energy and water on the planet. It occurs most strongly in the tropics, because the sun shines most strongly there.

A tropical cyclone is this process in the express lane. It is the atmosphere's special way of getting the heat out of the ocean more efficiently. The intense winds accelerate the ventilation of the ocean over a larger region, moving the heat from the ocean and into the atmosphere more quickly. The convection in the saturated eyewall moves the heat (much of it first held in the form of water vapor) upward without inhibition, bringing it as high as the stratosphere.

We know roughly how hurricanes function within the climate system. Can we predict how hurricanes will change as the climate system itself changes as a result of greenhouse warming?

There are some things we can say with confidence. Some other fundamental questions remain open.

Intensity—Reaching One's Potential

In the days before landfall, as Sandy moved north along the coast, it passed over a region around the Gulf Stream where the sea surface temperature, normally warmer than the water farther offshore, was yet still warmer than normal. The sea surface temperature "anomaly," or difference from what it would normally be at that time of year at any given place, was as much as a degree and a half Celsius—not a small difference to a hurricane. Simulations by the European Centre for Medium-Range Weather Forecasting after the fact, using the same model that produced the best forecasts, showed that the storm was more intense because of that sea surface temperature anomaly, with a minimum pressure 5 to 10 hectoPascals deeper than it would have been had SST been normal for the time of year.[96]

Much of the SST anomaly under Sandy's track was likely a result of natural variability, rather than caused by greenhouse gas warming. But some of the same physical understanding that we use to interpret the ECMWF result underlies our understanding of the likely influence of global warming on tropical cyclones.

One of the most important advances in dynamical meteorology of the twentieth century is the theory of the potential intensity of hurricanes. The primary originator of this theory is Kerry Emanuel of the Massachusetts Institute of Technology.[97] In the 1970s, Emanuel did his PhD under the supervision of Jule Charney, architect of the first successful numerical weather predictions. Emanuel's theory, developed in the 1980s, tells us that a hurricane is a heat engine.

The basic physics of heat engines has been understood since the nineteenth century. The essential idea is embodied in a conceptual model, called the Carnot cycle. It is named for the nineteenth-century French physicist Sadi Carnot. I visited Carnot's tomb while writing this book; he died in 1832, at the age of just thirty-six, and is now interred in the Panthéon in Paris.

A Carnot heat engine has a working fluid—in the simplest case, it's a pure gas—that undergoes a precisely defined set of four transformations, expanding and contracting, and cycling between two temperatures, a hot one and a cold one. Heat is added to the fluid when it's hot and taken out when it's cold. The net result is that some of the heat is transferred from the hot place to the cold place.

Heat flowing from hot places to cold places is the natural order of things;

the second law of thermodynamics, worked out in more generality by others after Carnot, enshrines that principle. But in a Carnot cycle, some of the heat gets siphoned off to do "work." Work means energy that can be used to do something you want to do. Drive a car, for example; an internal combustion system works on Carnot's principles.

Carnot's theory tells us what fraction of the heat moving through the system can possibly be turned into work. (The rest just goes to heating up the cold place, the same as if the hot and cold places had been put in contact without any engine in between.) That fraction, known as the Carnot efficiency, is the difference between the hot and cold temperatures (in degrees Kelvin, measured relative to absolute zero) divided by the hot one. The greater the temperature difference between the reservoirs, the more efficient the engine, all else being equal. Many other things can make a heat engine less efficient than the Carnot limit (friction in a car engine, for example). Nothing can make it more efficient.

The beauty of Emanuel's theory is that it makes the conceptual leaps necessary to apply the power and simplicity of Carnot's theory to a hurricane, which at first glance doesn't look like a simple heat engine. The warm temperature, in Emanuel's theory, is that at the sea surface, while the cold temperature is that at the top of the storm, where the eyewall convection flows out into the lower stratosphere. The heat input occurs in the form of evaporation of water into the surface air—remember, water vapor is heat, just in latent form—and the heat output occurs when the warm air at the top emits infrared radiation to space. The working fluid is moist air, with condensation of vapor into liquid occurring continuously in the eyewall. The hurricane transfers the heat from the warm ocean to the cold stratosphere and, in the process, turns some of the heat into work. The work is done on the ocean, as the wind generates currents. The same currents produce storm surge when the storm gets close to land.

In the simplest version of Emanuel's potential intensity theory, the efficiency is just that of Carnot: the temperature of the hot place minus that of the cold place, divided by that of the hot place. The sea surface is around 300 Kelvin, the stratosphere around 200, so the Carnot efficiency of a hurricane is one-third: $(300 - 200)/300 = 1/3$. More usefully, the theory gives a prediction of the maximum intensity that a hurricane can achieve, measured as a sustained surface wind speed, for a given climatic environment. The environment is specified in terms of the sea surface temperature, and the temperature profile as a function of height in the atmosphere. In our current climate, the theory predicts maximum potential intensities similar to what the strongest hurricanes actually achieve.

The environmental variables that control potential intensity in Emanuel's theory will change as the climate does. We can estimate these changes using modern climate models.

Today's climate models, for the most part, do not simulate hurricanes themselves very well. Their horizontal grid cells (like the pixel sizes in digital cameras) are much too large to allow narrow hurricane eyewalls to be resolved. But they can simulate changes in the climate itself, the environment within which a hurricane could form. Potential intensity theory allows us to estimate the maximum intensities of the hurricanes that we should expect in those virtual changed climates.

The results of such calculations unambiguously indicate that the maximum potential intensities of hurricanes will increase in warmer climates. This occurs because the sea surface temperature goes up, while the stratospheric temperature of the outflow does not. (It can actually cool; greenhouse gases do in fact cool the stratosphere as they warm the troposphere, although the strongest cooling occurs higher up than hurricanes can reach.) The Carnot efficiency of the storms increases slightly. The amount of heat they cycle through them, initially in the form of water vapor extracted from the sea, increases more, and can lead to significant increases in intensity.

How much hurricanes' maximum potential intensities increase in a warmer climate depends on the detailed structure of the climate change. How the warming is distributed over different heights in the atmosphere is particularly important. The temperature change just above and below the tropopause (the boundary between the troposphere, where we live, and the stratosphere, where the ozone layer is, starting around ten miles up in the tropics) is critical. Different models make different predictions about these quantitative details. But as with sea level rise, the *direction* of the change is consistent: models agree that the strongest hurricanes will get stronger.

While potential intensity theory gives us powerful insights into the basic physics of hurricanes, it has limitations. Some have to do with the validity of the theory's assumptions: a hurricane is not really isolated from its environment, as an ideal Carnot engine should be; its internal structure is not exactly what the theory assumes it is; it is not really circularly symmetric; and so forth.[98]

But even on its own terms, potential intensity theory doesn't explain the *actual* intensity of a hurricane, only its *potential* intensity—that is, the maximum possible intensity a storm can reach in a given environment. Most hurricanes, like most people, don't reach their potential. Wind shear, dry air, and their own imperfections (disturbances from circular symmetry) bring them down. None of those factors is directly included in the theory. It's possible that those factors could change in such a way that while maximum potential intensities go up, the intensities of real hurricanes don't.

That doesn't seem likely. In the current climate, it seems that the factors that limit real hurricanes act more or less randomly, on average, so that a hurricane has a roughly equal chance of achieving any intensity up to its maximum.[99] As long as that remains true, then increases in potential intensity will be accompanied by increases in real intensity. Even though most storms won't reach their potentials, on average they will come as close as they currently do, so increased potential will mean increased achievement.

Intensity—Models

We don't have to put all our trust in theory. As we so often do when studying (or forecasting) weather and climate, we can use computer simulations as experimental tools. In numerical models, we don't have to make all the simplifying assumptions that are necessary to achieve the clarity and simplicity of a theory such as Emanuel's.

Historically, though, the relationship of hurricanes to climate has been a difficult problem to address with numerical models, because of the wide scale gap. Normal climate models have grids too coarse to produce good hurricanes. Models with much finer resolution can produce much better ones, and have been used for decades to make hurricane forecasts, with some success. But the fine resolution makes those models much more computationally expensive to run. For hurricane forecasts, it is enough to simulate a small fraction of the planet for a few days. But to simulate climate, one needs to simulate the entire planet for many years. That has, until recently, been too expensive to do with resolutions fine enough to produce good hurricanes. One could produce a realistic virtual hurricane or a realistic virtual climate, but not both at the same time.

In the last decade or so, this has changed. Computers have become still more powerful, and models have been developed that can produce pretty good hurricanes, while at the same time it is possible (on the biggest supercomputers) to run them over the whole globe for decades. These are still not quite full climate models, in that they are "atmosphere-only"; it has still been too expensive to include organically simulated ocean currents. But by specifying the sea surface temperature field, one can impose a basic degree of control over the model's climate. The atmospheric temperature and wind will adjust globally to the sea surface temperature so that the two are broadly consistent. A sea surface temperature field taken from our current climate will produce weather very similar, statistically, to our own. Some of that weather will be hurricanes, emerging naturally.

A number of these high-resolution global models have been developed and used to study the virtual hurricanes' relation to climate. Perhaps the best at present is the one built by a group of scientists led by Isaac Held at the NOAA Geophysical Fluid Dynamics Laboratory (GFDL), in Princeton, New Jersey. This model, known as HiRAM (High-Resolution Atmospheric Model), produces truly amazing simulations of historical hurricanes in the Atlantic. When Held and colleagues tell the model what the sea surface temperatures were for each year in recent decades, the number of simulated hurricanes fluctuates year to year with remarkable fidelity to the real historical record. The model's grid is still too coarse for its hurricanes to achieve the peak intensities that the strongest storms achieve in the real world, but they come much closer than those produced by traditional climate models. The GFDL model can get up to about category three, whereas a normal climate model, such as those used to produce the warming projections for the most recent IPCC report, has trouble breaking category one.

To produce realistic global warming in HiRAM, the GFDL scientists increase both sea surface temperature and carbon dioxide at the same time.[100] In these warming simulations, HiRAM's hurricanes get more intense, just as potential intensity theory says they should. The model storms' intensities are still limited by resolution; on average, they are still too large and too weak, because of the large pixel size in the model. But they are otherwise very realistic, and not subject to any of the simplifying assumptions of potential intensity theory. They are not circularly symmetric about the storm center, and they are exposed to realistic wind shear and dry air. They emerge naturally out of the model physics, which is as faithful a representation of the real global atmospheric physics as can currently be produced on a computer.

Tom Knutson, one of Held's colleagues at GFDL, led a study in 2010 in which he and a group of scientists from several institutions compared results from a range of models, all of which were used to address the question of how hurricanes would change in a warmer climate.[101] Some, like the HiRAM, were global, high-resolution models. Knutson himself ran a still-higher-resolution model covering only the Atlantic, using input from other, global climate models to influence the simulated Atlantic climate. Kerry Emanuel used a "downscaling" technique, based on a hurricane model that is still simple but that admits more complexity than his potential intensity theory. The prediction that hurricane intensities increase in warmer climates was very consistent across all these models.

The intensity increases predicted by the study of Knutson and colleagues varied from 2 to 11 percent at the year 2100 relative to the present, depending on the model used. That may not sound large, but since damage increases proportionally to the cube or higher power of the wind speed, a 10 percent increase in intensity implies a 30 percent or greater increase in damage. And that was just the *average* intensity increase, over all storms. What matters most is the chance for a really strong one, and it's not clear that the models can predict yet how that will change. And of course, the models keep improving. The 2–11 percent number could just as well go up as down as the science evolves.

To summarize: as the climate warms, hurricanes will get more intense. We are pretty certain about this. Multiple lines of evidence support it: our best computer models, much better than any in the past, agree on it, and we have a good understanding of the physics, based on potential intensity theory. There is still uncertainty about how rapidly intensities will increase and about other details. But as with sea level rise, there is little doubt about the direction of the change; no one seriously argues that hurricanes will get weaker in a warmer climate.

Frequency—What We Know and What We Don't

Our understanding of intensity tells us something only about how strong a hurricane would be *if* it were there, not whether it is there in the first place.

Perhaps the most basic question we can ask about tropical cyclones' relationship to climate is how many storms a given climate will produce.

As the climate warms, this becomes a serious question. Since hurricanes form over warm water, and all the water will be warmer in a climate heated up by greenhouse gases, shouldn't there be more hurricanes in that warmer climate? More hurricanes, all of them more intense—is that what we are in for?

Or, what if hurricane intensities increase but there are far fewer of them? Can we rule that out? There would still be some risk of huge disasters from those few big ones, but wouldn't the overall risk go down?

Here, our science is in a more primitive state, compared to our understanding of intensity. Just in the last few years, we have started to get a grip on the relationship of hurricane frequency to climate, but we still can't answer these questions very confidently.

We do know a good deal about what kinds of climatic conditions are conducive to hurricane formation. Both forecasters and researchers have developed this knowledge over many years, starting in the early days of tropical meteorology. Some of it was described in 1948 by Erik Palmén, a Norwegian of the Bergen School.[102] The outlines of our present understanding were put more thoroughly in place in the late 1960s and early 1970s, by William Gray, a student of Herbert Riehl (himself a student at the Institute of Tropical Meteorology in Puerto Rico, taught by Clarence Palmer, put in place by Rossby). Later, Gray would use this understanding to produce the first seasonal forecasts of tropical cyclones—that is, forecasts not of a specific storm, but of the total number that would form in an entire season, the forecasts produced months in advance.

Hurricanes are prone to form over high sea surface temperatures; when the atmosphere is very humid; close to the equator but not right on it; and they don't like vertical wind shear. This understanding is built on experience, and confirmed by our historical data on past hurricanes and the conditions under which they formed, but it is not purely empirical. We do know enough about the physics of hurricanes to understand why these factors are important.

We do not, however, have a proper theory for hurricane frequency. We would like to have a set of logical arguments, based on solid physical principles and a set of clearly stated simplifying assumptions—something with the power and simplicity of potential intensity theory, with its foundations in Carnot's theory of heat engines—that would tell us how many hurricanes should occur per year in a given climate. We don't have it.

In our current climate, there are about ninety tropical cyclones per year distributed around the whole globe, plus or minus ten or so. Why this number and not double it, or half? We really don't know. Tropical meteorologists are notorious for vigorous disagreement, but in this case that's not the problem. No one has proposed any serious theory that would allow us to explain this number. Collectively, thus far, we have drawn a blank.

We do have a rough understanding of why the storms are distributed the way they are in space and time: why the western Pacific, for example, has more than the Atlantic (historically, close to thirty per year versus closer to ten); why the peak seasons are summer and fall in most hurricane-prone regions, whereas winters usually are hurricane-free; and why there are more eastern Pacific hurricanes, but fewer Atlantic ones, in an El Niño year. These things follow from our understanding of the factors that control tropical cyclogenesis (sea surface temperature, humidity, wind shear, rotation) and the ways that the climate varies from place to place, or from El Niño to La Niña.

But until we can explain the absolute number better, our predictions of changes in hurricane frequency with climate won't be as confident as we'd like them to be.

One thing we do understand very clearly is that the naïve argument based on increasing sea surface temperature—warmer sea surface temperature must lead to more hurricanes—is simply wrong. Hurricanes form over the warmest waters in the present climate, but it does not follow that increasing sea surface temperatures over the whole planet will lead to more hurricanes. Hurricanes form in places where the ocean is relatively warm *compared to the atmosphere*. As the climate changes, both the ocean and atmosphere temperatures will change in sync, confounding any simple link between the ocean temperature *alone* and tropical cyclones.

A hurricane forms from a cluster of normal tropical cumulus clouds. In their early stages, clouds that lead to a hurricane are no different from ones that don't. Air rises in cumuli when it is warmer than its environment. Updraft air that has started as surface air in contact with warmer water will be warmer (and moister, which will help warm it more once the water starts to condense) than if it had started over cooler water. But if the upper atmosphere through which the updraft rises is also warmer, the updraft needs that extra buoyancy just to keep going. On the other hand, if the atmosphere is cooler, the bar for how warm the updraft has to be to keep rising is lowered. The critical thing is the relative difference between updraft air and environmental air, and the updraft takes its properties from its initial contact with the ocean surface.

In a warmer climate, everything will warm—the whole ocean surface, the whole atmosphere—throughout the tropics. The air starting in clouds will be warmer and moister, but it will be rising through a warmer atmosphere. The warming of the atmosphere will just about keep pace with the warming of the oceans, so that the overall patterns of cumulus convection will stay similar.

This is not an accident. The cumulus clouds themselves do the job of taking the heat delivered to the ocean by the sun and bringing it into the upper atmosphere in the form of condensing water vapor. There are always many clouds at work doing this job, keeping the ocean and atmosphere in a closely coupled state. Any new cloud is rising into an atmosphere where many others have recently been—if not right in that place, then somewhere else, but the circulation spreads the warmth out over the whole tropics. (The warming is confined mostly to the tropics, because at higher latitudes, the Coriolis force becomes stronger, deflecting air currents sideways and keeping them trapped closer to their points of origin.)

Climate dynamicist J. David Neelin at the University of California, Los Angeles, coined an analogy that compares the tropical atmosphere to a poker game in which the entire tropics is playing. Each cumulus cloud anywhere in the tropics "ups the ante" for all the others by warming the atmosphere tropics-wide.[103] A hurricane, the most intense and organized system of cumulus clouds, needs to be able not just to meet the ante, but to raise it still more.

As a consequence, hurricane formation will still be limited to the warmest regions of an overall warmer climate. The threshold sea surface temperature (the degree of warmth necessary to form a hurricane) will have risen, because clouds trying to form a hurricane will be rising through a warmer atmosphere, requiring more initial warmth and moisture from the ocean below. The cooler patches in the new climate, though warmer than they were before and even warmer than what the threshold used to be, will stay below the new threshold.

That's not to say that the patterns of hurricane formation won't change at all, or that their total number won't change. But this argument is wrong: "right now it takes water at eighty degrees Fahrenheit to form a hurricane; in a climate that is two degrees warmer everywhere (say), a larger area will have sea surface temperature over eighty degrees; therefore the area where hurricane formation is possible will increase, and there will be more hurricanes." It is wrong because if the climate warms two degrees, the threshold for hurricane formation will also go up, probably also by about two degrees.

Frequency—Models and the Ballooning Deficit
In the absence of theory, we have to rely more heavily on numerical models. Our thinking on how hurricane frequency will change in a warmer climate has evolved rapidly in the presence of the new global high-resolution hurricane models, such as the HiRAM from GFDL. For our understanding of intensity, those models are an important source of information; for frequency, they are critical. Those models broadly agree that in a warmer climate, the average annual total number of hurricanes on the earth will decrease. For the most part, so far, the models say that a warmer world will have fewer hurricanes, not more.[104]

That result was counterintuitive, at first, a few years ago. Many of us still had the intuition that warmer oceans should produce more hurricanes, even if we knew that the most naïve arguments to back up that intuition were wrong. We are starting to understand it now. It seems to be a subtle consequence of the same physics that controls the water vapor feedback, the same process that makes global warming as powerful as it is.

As air temperature increases, the saturation specific humidity (the maximum amount of water vapor that can be in the air without condensing) increases. All our climate models, and several other kinds of evidence as well, suggest that the relative humidity stays close to constant. The relative humidity is the fraction of the saturation value: how much water vapor is actually in the air, divided by how much there could be. So if it stays constant, while the saturation value increases because of warming, the actual amount of water vapor in the air must increase. This is what causes the positive water vapor feedback, since water vapor is a greenhouse gas.

It seems, though, that hurricane formation is sensitive to what we call the saturation deficit. That is the absolute difference between the actual humidity and what it would be at saturation. It's almost the same as the relative humidity, but not quite. It's the difference, not the ratio, between the actual and saturation values. It is how much water vapor (the actual amount, not the percentage) one would have to add to make the atmosphere saturated. As cumulus clouds try to get organized into a hurricane eyewall, they can be more easily hindered if that difference is larger.[105] But if the relative humidity is the same while the total humidity increases, this implies that the saturation deficit goes up.

The math is simple, but subtle at the same time. A typical saturation specific humidity for air near the ocean surface in the tropics could be 30 grams per kilogram. (The water vapor makes up 3 percent of the air by mass.[106]) Let's say the relative humidity is 80 percent; that means the actual specific humidity is 24 grams per kilogram. The saturation specific humidity is the difference, 30 − 24 = 6 grams per kilogram. Now let's say global warming occurs, and with the new higher temperature, the saturation specific humidity is now 35 grams per kilogram. (That's a reasonable ballpark estimate for the warming we expect sometime in the middle of the twenty-first century relative to the late twentieth.) If the relative humidity stays at 80 percent, the specific humidity will be 28 grams per kilogram. The saturation deficit, then, is 35 − 28 = 7 grams per kilogram, one gram per kilogram greater than before. That increased deficit means, it turns out, that it's harder for a hurricane to form. Cumulus updrafts are now 7 grams per kilogram drier than their environment instead of 6, and that slows them down more as they mix with the clear air outside. Once a hurricane actually forms, it doesn't matter as much (if there is not too much shear), because the eyewall becomes saturated and its updrafts insulated from the air outside the storm. But that extra saturation deficit matters when a new storm is trying to get going.

The full story, though, is more complicated than just humidity. All the factors that influence hurricane formation will change in a warmer climate. Wind shear, for example, will get stronger in some places, weaker in others, as the trade winds and jet streams shift around.

The detailed patterns of ocean temperature change matter as well. Hurricanes will form over the warmest oceans, wherever those are; if those become warmer *relative* to the rest of the oceans, then more hurricanes will form over those hot spots. It's possible that increases in hurricane frequency in those hot spots could outpace decreases elsewhere to add up to an increase. But the models seem to show that the downward pressure from the saturation deficit wins, because it's global in scale. It happens basically everywhere.

Different models, though, show reductions in hurricane frequency that are quite different in degree. Some show large reductions; others show much smaller ones. And Kerry Emanuel's downscaling technique, using a simple but elegant model to produce hurricanes based on climatic environments taken from global climate models, has now broken the consensus. When used with the simulations from the previous generation of climate models (the ones used to produce the fourth IPCC report in 2007), Emanuel's technique produced a decrease in hurricane frequency, like the global high-resolution models; now, with the latest generation of simulations run for the fifth IPCC report, it produces an increase.[107] I don't think we fully understand this result yet. But as long as we have to rely so heavily on numerical simulations to tell us how many tropical cyclones will form per year in the future, we have to take every credible simulation seriously, and Emanuel's is certainly credible.

Intensity Versus Frequency: The Many and the Few

So, we are quite confident, based on both a basic understanding of the physics and our best computer models, that tropical cyclones will be more intense in a warmer climate. We are less confident about whether there will be more of them or fewer. Lacking a deep understanding, we rely more heavily on the models. Those have been telling us, mostly, that there will be fewer tropical cyclones in the future rather than more. Let's accept this for the sake of argument.

If hurricanes are more intense but less frequent, what does that mean for our risk of a major disaster? By far most of the damage inflicted by tropical cyclones comes from the most intense ones, even though those are a small fraction of the total number of storms. A category three doesn't sound that much worse than a category two, but the damage escalates very rapidly with each step up the Saffir-Simpson scale. A category four is much worse still. A category four is very unlikely to hit the northeastern United States anytime in the near future, but if one were to do so it would be unlike anything ever experienced in the region. It is worth asking whether the risk of such an intense storm—or, let's say, at least a category three, which is rare but known to be possible even up here, because one hit Long Island in 1938—will go up or down.

Whether there are more or fewer of the most intense storms depends on which of the changes—more intense or less frequent—wins out. Imagine you have a group of athletes trying to jump over a high bar. Unassisted, a few of them can do it. Then they all take performance-enhancing drugs. They can all jump higher than they could before. But at the same time some of them drop out of the competition; let's say they got caught by mandatory drug testing. Does the total number of athletes clearing the bar go up or down? It depends on just how many drop out versus how big a performance boost the remaining ones get. With hurricanes, we don't know either of the changes with enough precision to see which one will dominate. The precautionary principle, though, suggests that we should be concerned about an increase in risk. There is certainly a significant chance, based on the best current science, that the global frequency of major hurricanes (categories three and up) will increase in a warmer climate.

Absolute and Relative

If we are concerned not about the whole world's hurricane-prone coasts, but just a particular region, the factors we have to consider change somewhat. To the limited extent that there is a model consensus on a decrease in tropical cyclone frequency, it is at the global level. It falls apart completely when we consider any of the individual regions, or "basins," in which the storms form. This is a consequence of limitations in our ability to predict the detailed patterns of regional climate change, perhaps even more than it is a consequence of our limited understanding of tropical cyclones.

While some of the effects of global warming on tropical cyclones are subtle, the effects of *local* warming are simpler. If some part of the tropical ocean warms more than the rest, the warmer part will become more favorable to tropical cyclones. If some other part warms less than the rest, it will become less favorable to tropical cyclones, despite being warmer than it used to be.

Starting in the 1970s, the North Atlantic, in particular, warmed more than average for about thirty years. Part of this excess warming may have been a consequence of natural fluctuations in the deep ocean circulation. At least some of it, and maybe most, was probably due to reductions in sulfur emissions in the United States due to the Clean Air Act.

Sulfur gases emitted from midwestern smokestacks formed tiny aerosol particles that dimmed the sunlight at the surface, cooling the Atlantic as the midlatitude westerlies blew the particles out over it in the 1950s, '60s, and '70s. When the new emission restrictions thinned that aerosol cloud in the 1980s and after, the Atlantic warmed rapidly, both in absolute terms and compared to the rest of the global oceans—though the rest were also warming, in large part due to carbon dioxide.

During the same period, the frequency of tropical cyclones increased. In the 1970s and '80s, when the Atlantic was at its relative coolest, there had been few storms. The numbers climbed in the 1990s and 2000s, peaking in the 2005 season when Katrina, Rita, and all the rest used up our Roman alphabet, forcing NHC to dig into Greek letters to name the storms as the season ran, unprecedentedly, into January 2006.[108]

The correspondence between the increasing sea surface temperature—in the Atlantic, but also globally, as both were rising (the Atlantic, just more)—and the rapidly rising number of hurricanes was striking. It was tempting to draw the conclusion that continued global warming, by increasing sea surface temperature further, would threaten catastrophic increases in hurricane activity. With remarkable timing, Kerry Emanuel published a paper during the summer of 2005 suggesting this.[109] Within weeks, a team led by Peter Webster, another extremely prominent tropical meteorologist, published a paper drawing similar conclusions on a global basis. Webster and colleagues' paper indicated a large increase not in the total number of storms, but in the number of category four and five storms, the most destructive by far.[110]

The Emanuel and Webster papers, coming when they did in the middle of the most extreme Atlantic hurricane season ever by far, fueled enormous debate. The media suddenly paid much more attention to the question of how climate change would affect hurricanes.

So did scientists. The number working on the problem jumped in the space of a few months. Before that, climate science had been a big field, and hurricane science had been a big field, but the overlap between them had been small. Most climate scientists didn't know much about hurricanes, and most hurricane experts didn't like to think about climate. Both began to change quickly in 2005.

Suzana Camargo had been working at the International Research Institute for Climate and Society (IRI) at Columbia University on developing a method for making seasonal forecasts of tropical cyclone activity (forecasts of an upcoming season as a whole, rather than of single storms). As an outgrowth of her work, she and I had begun doing some more basic research together on the relationship of tropical cyclones to El Niño. A few months before the 2005 Atlantic season, we and a couple of other colleagues began organizing a workshop on tropical cyclones and climate. Both being relatively new to this particular research area, we wanted to make stronger connections with others working in it. We expected a small, quiet workshop.

Then the 2005 season happened. By the time we held our workshop, in March 2006, we felt we were in a hurricane ourselves. Attendance was much higher than we had expected. The changes of the last few decades in the Atlantic were the subject of much of the discussion, some of it quite animated.

In the years since then, it has become clear that the Atlantic story is not simple. Given what we know about tropical cyclone formation, it seems reasonable to interpret the rapid rise in Atlantic hurricane frequency since the 1970s as a result of the *relative* warming of the Atlantic, rather than its *absolute* warming.[111] Whether the relative warming resulted from ocean circulation or the Clean Air Act, there is no reason to expect it to continue. Ocean circulation fluctuations reverse eventually, and the heavy aerosol pollution of the mid-twentieth century can't be removed again, since it's already gone. Our default expectation should be that the Atlantic will warm at about the same rate as the rest of the tropical oceans.

Even slight differences in climate from one place to another matter, though. While the warming due to greenhouse gases is indeed global, it is not perfectly uniform. It has patterns. When we look at regional patterns of climate change that have already occurred, such as the recent Atlantic warming, it is very difficult to separate out the different possible causes. But in climate models, we can change greenhouse gases only, keeping all other factors the same. When we do this, there are some patterns that emerge consistently in many models. Barring other influences that are difficult to foresee—a Clean Air Act in China?—these are patterns we might expect to be most likely to emerge in the future, as greenhouse warming strengthens relative to natural variability.

In most climate simulations driven by greenhouse gas increases, the

Northern Hemisphere, including the Atlantic, warms a little more than the Southern.[112] This could increase Atlantic hurricane activity.

At the same time, there is a band of water stretching across the Caribbean (prime territory for cyclogenesis, at present) that warms less than average.[113] This could reduce Atlantic hurricane activity, since it will be a cool spot in a relative sense. It's cool partly because of strong trade winds blowing over the ocean and driving strong evaporation of water vapor from the surface. This, in particular, is a factor that suppresses genesis.[114]

The models also predict, very consistently, that the eastern and central equatorial Pacific will warm more than the rest of the ocean. This produces a pattern similar to what happens in an El Niño event. El Niño events suppress Atlantic hurricanes, partly because they shift the jet stream south over the Atlantic, increasing wind shear there. The models produce the same shear pattern in response to greenhouse warming.[115] At the same time, an El Niño encourages eastern Pacific hurricanes and strengthens western Pacific typhoons. So we might expect a future climate to have more and stronger storms in the Pacific but fewer in the Atlantic.

Despite the consistency of the broadly El Nino–like warming pattern across models, though, we are still not entirely certain it's correct. The models' simulations of the current climate in the eastern and central Pacific are flawed. The El Niño–like pattern results from a competition between different processes in the atmosphere and the ocean;[116] some of them would tend to produce the opposite, La Niña–like state, with less shear and more hurricanes in the Atlantic. If those La Niña–inducing processes are a little too weak in the models or the processes favoring the El Niño–like state are too strong, the entire answer could be wrong as far as the patterns of hurricane change are concerned. This is the frontier of our current understanding of those patterns.

Size Matters; Transition

Nearly all our research on the relationship between tropical cyclones and climate addresses statistics. Even when looking at extreme events, we think about averages. We have to. While each event is different, it's too hard to think about all the possible types of storms at once; if we want to get anywhere, we have to focus on what's common.

"Hurricanes typically form under such conditions, or don't, with such probabilities." "Such factors are, on average, favorable or unfavorable for a hurricane to become more intense, by such a margin." We have to try to construct a rigorous scientific basis for statements of this type.

Then we get an event that looks different from any other we have ever observed. What can we say about how that event was related to the climate in which it formed? What do averages have to do with it?

The most intense hurricanes typically do the most damage. Sandy, on the other hand, did a huge amount of damage despite being close to the bottom of the current Saffir-Simpson scale, which measures only wind. Its winds were those of a low category one hurricane at landfall. It was so destructive (with a surge that would rank in category three on the old Saffir-Simpson scale) because of its enormous size, and because it came in on a westward track, pointing strong winds over a huge fetch into New Jersey and New York Harbor. How were these factors related to climate? Does global warming make hurricanes larger, or more likely to travel westward?

We know rather little about how the size of tropical cyclones is related to climate. Reliable observations of storm size go back just a couple of decades, and don't yet show any clear changes in time. If we want to try to understand how a warming climate will affect storm size, we will probably need numerical models. But storm size is a weak point of even the best current global high-resolution models, such as NOAA's HiRAM. In order to be able to run them long enough to study climate change, these models need grid sizes that are still a little too large to resolve tropical cyclones fully. They may get the right numbers of storms, but their storms are still uniformly too large compared to real ones.

Sandy became as large as it did because it was a hybrid. In fact, this is how many of the largest storms are formed.[117] The process of extratropical transition, typically occurring as a tropical cyclone moves poleward, usually broadens its wind field. Extratropical baroclinic cyclones tend to be larger than hurricanes, as a consequence of the large scale of the jet stream insta bility that forms them. As a hurricane draws energy from the jet stream like an extratropical storm, it grows to match that large scale.

So how will the process of extratropical transition change in a warmer climate? Will more tropical cyclones make it far enough poleward to become hybrid storms, or fewer? Will those that do become stronger or weaker?

These questions have been studied almost not at all. Extratropical transition has not been viewed as the most-high-impact area of tropical cyclone research. The process most commonly occurs over the open ocean, and many storms that make the transition weaken rather than strengthen as they do, which further diminishes them as threats.

Sandy, on the other hand, transitioned dramatically, strengthened during the process, and then made landfall. But that was, by historical standards, a freak occurrence. The incapacity of NOAA's warning protocols to handle it made that plain. Up to now, an event like this had not seemed nearly as probable a threat as a more typical hurricane landfall. Not only did the rules about storm warnings reflect that, but the efforts of researchers did as well. As a consequence, there has been essentially no research on the relationship of extratropical transition to climate.

Now, post-Sandy, there surely will be new research that will try to assess how the frequency, intensity, and tracks of storms undergoing extratropical transition will change in the future. My colleagues and I will do some of it ourselves. But so far, there is little we can say about how climate change may influence this process that so thoroughly defined Sandy.

Track Changes

When I first saw the left turn in the model predictions for Sandy, I didn't realize how unprecedented it was. My memory of all the past historical hurricane tracks in the Atlantic wasn't good enough for me to be sure there hadn't been one or two like that. By the day after landfall, I (like many others) realized that the left turn and westward-moving landfall were unique in the last 150 years, since the beginning of the Atlantic hurricane record. This, just as much as the size, made Sandy a profound outlier, and likely contributed as much to the catastrophic storm surge. Its not having been observed in 150 years clearly made the westward track rare, but how rare? What was the "return period," the typical time between such events one would find if one had an infinitely long historical record? Was it a hundred-year, thousand-year, or ten-thousand-year storm track? Just how unlucky were we?

My colleague Tim Hall at the NASA Goddard Institute for Space Studies has built a mathematical model that is good for questions like these. It is not a dynamical model, like those used to predict the weather or simulate climate change; it does not solve the physical equations of motion. It is a statistical model that takes what we call a "Monte Carlo" approach to hurricane risk. Tim's model generates artificial hurricane tracks that have the same statistical properties as real ones. He then uses those artificial tracks to estimate the probabilities of events that have never occurred. Tim originally wrote this model for the insurance industry, which is keenly interested in the probabilities of rare but disastrous events.

Because Tim can very easily generate a huge number of storms, he can produce a much larger "historical" database than we have for the real atmosphere. This database will have some flukes in it—for example, storms that are of such rare types that we have never seen them, but that, in a meaningful statistical sense, are still consistent with the ones we have seen. By generating enough storms of some specific rare type, Tim can estimate just how rare a real storm of that type should be.

On Tuesday, October 30, the morning after landfall, I was in the midst of media interviews. Many of the interviewers asked questions about the rareness of the storm. In thinking about this, I quickly realized that Tim's model would be a good tool to estimate the probability of a Sandy-like track. When I contacted him about it, he was immediately interested, and got to work on a set of calculations that we turned into a research article a few months later.

Tim used his model to estimate the return period for a storm of at least category one intensity making landfall on the New Jersey coastline at an angle at least as steep as Sandy's. He came up with a value of seven hundred years, with an uncertainty such that we couldn't rule out values between four hundred and fourteen hundred years. That's a wide range, but a rare event no matter what.

That doesn't mean that the return period for the *flood* is that long. Consider New York City, for example. Sandy's track was one of the factors producing the storm surge here, but one could get the same surge from a storm coming in at a shallower angle and farther north in New Jersey (closer to New York City) or at higher intensity. My guess is that the return period for the flood is probably shorter than seven hundred years—maybe as short as a couple hundred. The historical record of flooding in the city is only a few hundred years long, and only since 1920 has the modern tide gauge at the Battery been in place to take really consistent data, but the record does suggest that Sandy's storm tide beat anything in recorded history, going back to colonial times. The major hurricane landfall near New York in 1821 may have produced a storm surge a little larger than Sandy's. But the peak surge came at low tide, and the resulting flooding in Manhattan is believed to have been just a little less than what Sandy caused. Of course, it wasn't the same city then; there were many fewer people, and no subways to flood.

In any case, the model Tim used to produce his estimate of a seven-hundred-year return period for the westward track doesn't account for climate change. It's an estimate based on the assumption that the climate is that of the last one hundred fifty years. But the climate is changing rapidly, and it's reasonable to think that the properties of storm tracks will change with it. If we want to know whether westward-moving storms will become more frequent or less in the future's warmer climate, we have to use other tools.

A couple of papers written in the months before Sandy argued that the melting of Arctic sea ice in recent decades had caused the jet stream to become more disturbed.[118] After Sandy, these papers were seized on by some in the media as implying that tracks like Sandy's would become more common in the future. Sea ice melting is an almost certain consequence of global warming (even if each individual year's sea ice also reflects natural fluctuations). So as the climate warms and more ice melts, these papers implied, we should see a more disturbed jet stream. If correct, this argument might lead one to expect more jet stream disturbances similar to those that were in place during Sandy, with blocking highs in the Atlantic and troughs over the eastern United States. This implies a more frequent occurrence of westward steering flows such as the one Sandy was caught in (the opposite of the normal flow at our latitudes, which is eastward), and greater likelihood of a westward track if a hurricane should drift into those westward flows.

My colleagues and I believe this conclusion is unjustified. Elizabeth Barnes, Lorenzo Polvani, and I published a paper to this effect in the summer of 2013.[119] Climate models very consistently show essentially the opposite kinds of changes to the jet stream as the climate warms. They show the jet stream shifting poleward; in the fall, this turns out to mean that the flow is actually more persistently eastward around New York and at higher latitudes, so that westward steering is less frequent, not more. They also show reductions in the frequency of blocking high-pressure systems. Similar results had been found in earlier studies by Chunzai Wang and colleagues at NOAA.[120]

The models are far from perfect. They don't produce blocking as often as they should in the current climate, so we can't be entirely certain they are right about the changes in the future. But if they are wrong, then all of them are wrong at once. The changes in the extratropical flow they depict are among the most consistent results to come out of the last decade of climate change research, from many scientists using many models. And the models allow one to separate cleanly what is caused by greenhouse gases from what is not, by doing controlled experiments and running them long enough to average out natural variability. We believe that the conclusion that melting sea ice leads to a more disturbed jet stream was based on an overinterpretation of a short record containing natural variability.

Natural Variability

Natural variability is not just a problem for interpreting the recent changes in blocking and assessing the influence of warming on Sandy's track. It is a fact of much broader importance as we try to understand tropical cyclone risk.

Understanding the many dimensions of global warming's impact on tropical cyclone risk is critically important for the long-term future. But if we want to understand hurricane disasters past, present, and near future, it's a sidebar. When we look at tropical cyclones in particular, the signal of climate change has not yet clearly risen above the noise of natural variability.

Everything in the climate system goes through natural cycles. Some cycles—for example, those that result directly from the orbits of the earth and other planetary bodies—are truly regular and predictable.

The march of the seasons, a consequence of the earth's orbit around the sun, is the annual cycle. The alternation of day and night, due to the earth's rotation about its axis, is the diurnal cycle. The moon's orbit drives the tides so predictably that tide tables can be published years ahead of time.

Over tens of thousands, even hundreds of thousands, of years, the gravitational pulls of the other planets cause regular cycles in the geometry of the earth's orbit around the sun. The average tilt of the earth's axis, for example, fluctuates back and forth every 41,000 years, while the axis itself rotates (or "precesses") around that average tilt every 26,000. The elliptical orbit of the earth stretches out, deviating a little more from a circular shape, and then squashes back to an almost perfect circle, every 413,000 years. These

are called Milankovitch cycles, after the Serbian scientist who first postu-
lated that they might influence the climate by changing the way sunlight is
distributed over the globe and through the seasons.

Indeed, as they go in and out of phase over geological time, the Milankovitch cycles drive ice ages and warm periods that paleoclimatologists can
read in the rocks. The ice ages that have actually been found in the geologic record, though, don't match the Milankovitch cycles in very precise or
simple ways. This is one indication that while external factors certainly influence the climate, the climate also has some freedom to do what it wants.

Besides the regular cycles of the orbits of the planets, the turbulent
weather in the earth's atmosphere and in the ocean, too, generates its own
cycles. The word *cycles* is somewhat misleading here, as these natural variations are nowhere near as regular or predictable as the seasons or the tides.
But when these natural variations make a large excursion in one direction,
they will eventually go back in the other direction, even if we can't say exactly when.

If we focus on time scales of modern human history (the last hundred
years, say), the Milankovitch cycles are irrelevant. They are just too glacially
slow. The earth's orbit is, for all practical purposes, the same at the beginning of any single century as at the end. Warming due to human greenhouse gas emissions is, frighteningly, much faster. Its signal is now quite
clear in the history of the global mean surface temperature of the last hundred years. Since the industrial revolution, the pollution we have put into
the atmosphere has warmed the planet by around 0.85 degrees Celsius, or
1.5 degrees Fahrenheit.[121]

But on top of that steady rise are large fluctuations. Some of these may
be anthropogenic as well; some may be due to fluctuations, for example,
in sulfate aerosols from industrial emissions. These may have suppressed
Atlantic hurricanes in the 1970s, as the aerosol cloud cooled the Atlantic
Ocean, and then encouraged them in the 1990s and 2000s, as the dimin-
ishing of the aerosol cloud allowed the ocean to warm up in the sun.[122] But
much of it is certainly natural.

Some variability has no cyclic character or particularly identifiable traits
at all. Some of what the climate system does is truly just noise.

But above that noise, we can identify some useful signals. We can define
a number of distinct types, or "modes," of natural climate variability. Each

has its own structure, often growing out of the particular geography of one particular region of the world. Each has its own time scale; some tend to stay in the same phase for decades, others just for weeks. Some arise mainly out of the atmosphere, others mainly out of the ocean, though all, ultimately, are influenced by both. Some are more predictable than others.

The Atlantic Ocean in particular seems to undergo slow, deep oscillations. The thermohaline circulation, in which cold, salty Arctic waters sink to near the bottom, meander south at depth to the opposite hemisphere, and slowly mix back up to the surface—a full circuit would take a time measured in millennia, if you could mark a bit of water and follow it through —strengthens and weakens, gradually, as decades and centuries pass. These somewhat regular, somewhat irregular cycles result in part from the fast, turbulent atmospheric weather pushing on the much heavier, slower-moving ocean. At the same time, the ocean itself has its own "weather," with disturbances in its currents that evolve more gradually, but perhaps ultimately just as chaotically, as the atmospheric winds do. The resulting fluctuations of the Atlantic Multidecadal Oscillation, or AMO, make the North Atlantic's surface warm or cool relative to that in the south, influencing hurricanes as well as other weather patterns.

Some of the slow variations in Atlantic hurricane activity (the quiet 1970s and 1980s, the active 1990s and 2000s) may be due to the AMO. As I mentioned earlier, another possibility is that these variations were not natural, but were caused by human activity—though not greenhouse gases. Pollution by clouds of tiny aerosol particles, emitted by industrial smokestacks in the eastern United States in the postwar period and then transported over the Atlantic by westerly winds, is believed to have cooled the eastern Atlantic by shading its surface during the 1970s and '80s. Being slightly cooler relative to the surrounding oceans, the Atlantic would then have been less favorable to hurricane development. When that pollution was reduced by the Clean Air Act and its various amendments (phased in from the 1960s to 1990), the sunlight increased again at the surface, allowing the Atlantic to warm and the hurricanes to come back.[123] The relative importance of this human influence via aerosol pollution versus the natural influence of the AMO remains a subject of disagreement among climate scientists.

In the quicker variations from one year to the next, El Niño dominates.

Properly, we talk about the El Niño–Southern Oscillation, or ENSO, phenomenon. At its core, ENSO occurs in the equatorial Pacific Ocean, but it influences the climate of the whole earth.

Near the equator, the surface of the western Pacific is warm, with frequent deep convection (tall, vigorous cumulus clouds), heavy rains, and typhoons in the overlying atmosphere. The eastern Pacific is cool, as easterly (westward) winds drive upwelling that brings cold water to the surface. As the upper atmosphere over the east is nearly as warm as that over the west, the much cooler surface air in the east is too dense and heavy to rise through it. Tall cumulus clouds and heavy rain are rare. The east normally has gray weather with low stratus clouds; this is what Darwin saw when he went to the Galapagos.

In an El Niño event, the upwelling weakens and the east warms, while the west changes little. The rainy weather in the west shifts eastward with the warming sea surface temperature. But this weakens the easterly winds, which themselves were a product of the normal east-west weather contrast; since the easterlies drove the upwelling, their weakening strengthens the eastern ocean warming further. These positive feedbacks—first described clearly in 1969 by Jakob Bjerknes, Vilhelm's son, who by that time worked at UCLA, in California[124]—can persist for many months. Eventually, changes in the deeper ocean currents ripple back and forth across the Pacific and disrupt the pattern. They sometimes reverse it entirely, making the east colder than normal and confining the rainy weather even farther to the west. That is called a La Niña, the opposite of an El Niño.

Though focused in the Pacific, ENSO has global impacts. We can see it even in the global mean temperature. El Niño events bring heat out of the ocean and into the atmosphere, and increase the temperature of the planet, temporarily, by as much as one degree Celsius (two degrees Fahrenheit). By bringing the heat into one specific place (the eastern and central tropical Pacific), ENSO also disturbs the atmospheric circulation. This disturbance has impacts on the climate of specific places much stronger than that which temporary global warming alone would have. El Niño events typically cause droughts in many tropical and subtropical places: Indonesia, Australia, southern Africa. They cause floods in others, such as the southwestern United States. La Niñas have effects that are approximately, though not exactly, opposite. The persistent drought of the central and southwestern United States in the last fifteen years or so appears to be attributable to frequent La Niña conditions.[125]

ENSO influences hurricanes globally. An El Niño event increases hurricane activity in the eastern Pacific, shifts it eastward in the western North Pacific, and northward in the South Pacific. In the North Atlantic, El Niño suppresses hurricanes. The jet stream shifts southward, putting more of the tropical and subtropical hurricane formation region into the strong wind shear under its equatorward flank. A La Niña does the opposite, and La Niña years often see active Atlantic hurricane seasons.

There are just a handful of modes of low-frequency climate variability that are coherent and regular enough to be somewhat predictable, or at least to explain seasonal variations in hurricane activity in a meaningful way. There are also higher-frequency fluctuations that modulate weather within a season. The Madden-Julian Oscillation is the most important of these.

Then there are modes of variability that have distinct, identifiable patterns that recur again and again, but with no particular regularity or predictability. The North Atlantic Oscillation, which shifts the jet stream north and south over the high-latitude Atlantic, is a good example of that. It is somewhat misnamed, because it doesn't oscillate in any regular way. It can flop back and forth quickly, or stay in one phase for weeks. It can even exhibit persistent shifts in its average state over decades, though still varying rapidly about that average.

Sometimes natural modes of variability come together in ways that just happen to add up to something big. Sandy grew out of such a confluence. While the question of how greenhouse warming influenced Sandy is important, the core of what happened was clearly natural. An MJO event passed through the Caribbean at the same time as a strong negative NAO event was pushing the jet stream far south over the eastern United States, and a blocking high (somewhat reminiscent of the typical NAO pattern, but farther south) was setting up over the western North Atlantic. A tropical cyclone grew out of the favorable MJO phase and made its way up the coast. Anomalously warm sea surface temperature, mostly natural in origin, likely kept its hurricane core going longer than it would have. The storm both grew to enormous size and turned left into the coast due to interactions with midlatitude upper troughs of the type that are favored in negative NAO events. The NAO and the ocean set up the pins, the MJO rolled the ball into them, and the exact timings of both the hurricane and the troughs added up to a strike.

Though this sequence of events was very predictable on a time scale of a week (apparently), such sequences are inherently unpredictable much farther in advance than that, due to the chaos of the atmosphere. The MJO and NAO fluctuations were short term; there was no La Niña or other clear favorable low-frequency shift in the Atlantic.

When we step back from Sandy, or any other individual storm, natural variability still makes it hard for us to see climate change signals in the recent history of hurricanes. Globally, and even more so in individual basins, hurricane activity varies from year to year and decade to decade. Most of that variation is natural. Some of it is due to coherent, low-frequency climate signals with roots in the ocean, such as the AMO and ENSO. Some of it is due to the random coincidences of chaotic weather in the atmosphere. This background of constant natural variability is like noise that, for the time being (and probably for the next few decades at least), drowns out the signal of any slow trend due to global warming that may be there—even if that signal is large enough to be very dangerous in the long term.

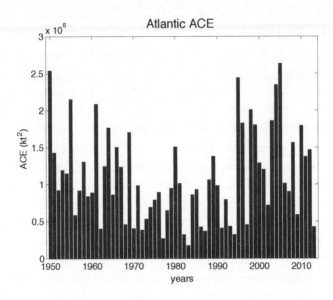

Time series of Atlantic Hurricane activity since 1950. Each bar shows the total accumulated cyclone energy (ACE) for the Atlantic Basin for the year indicated on the x axis. ACE is an integrated measure that includes the numbers, intensities, and lifetimes of all tropical cyclones in the North Atlantic Basin each year. (It is computed by taking the square of the maximum wind speed of each storm for each six-hour period of its life and adding up those numbers for all the six-hour periods for all storms.) *(Courtesy of Suzana Camargo, data from National Hurricane Center/NOAA)*

Seeing the global warming signal above the noise of natural variability is an issue in all climate science. It's an issue even when we look at the global mean temperature, the most basic measure of climate.

The global mean surface temperature has been flat for the last fifteen years; this is the so-called warming "hiatus," or "pause." Climate change deniers have become excited by this, seizing on it as evidence either that warming is not real or that climate scientists don't know as much as we say we do. They are right that we don't understand the specific causes of the hiatus as thoroughly as we would like to, but entirely wrong on the big picture.

Our knowledge of climate physics tells us that greenhouse gases warm the climate over the long term, but that we should also expect variability about that long-term trend, and that much of that variability will not be predictable. We see all this in the observations. The long-term trend in the temperature is clear, and the hiatus doesn't change it. It's like driving a car on a winding road through a hilly forest. Your map tells you the road leads off a cliff, and much of the time you can see the cliff ahead of you. But sometimes you can't see it; trees or hills block the view. Sometimes, when the road takes a big turn, you may even be temporarily driving away from it. Do you really want to let each turn make you doubt what both the map and your own eyes tell you about your overall direction? Do you just keep driving, or do you consider turning around?

When we look at hurricanes, the problem is more complex. Not only is our understanding of hurricanes' relation to climate less solid than our understanding of global mean surface temperature, but the natural variability is much larger relative to any long-term trend. If we look over a period of a few decades, we may see an upward trend in Atlantic hurricanes, but if we choose a different period, we see no trend or a downward trend. Even if warming is causing significant changes in hurricanes (the increases in intensity, for example, that we expect), it is too hard to see those slow, long-term signals above the noise of natural variability.

If we try to defeat this problem by looking over the longest periods we have, back to 1900 or even 1851, when the modern record of Atlantic hurricanes begins, we have the problem that the quality of the data at the beginning and the end is not comparable. Since the late twentieth century, we have satellites, and many other ways to see and measure hurricanes that we didn't have earlier. The farther back in time we go, the likelier it is that there were storms that we would have observed if they had occurred today but that simply went unrecorded at the time. Even those that were observed were not measured as accurately as they would have been today; their intensities, in particular, become more uncertain as we go farther back.

This is why, when we talk about changes to hurricanes as a result of global warming, we rely so much on theory and models. Unlike with global mean temperature, where the long-term trend is apparent to the naked eye, we can't clearly see the imprint of warming on hurricanes just by looking at the historical record of hurricanes from the past to the present. By no means should we draw the conclusion from this that the hurricanes are not changing due to climate change. The variability is just so much larger in the short term that, even if the trend is substantial, we won't clearly be able to detect it until hurricanes have become so much stronger that we will be in big trouble. Compared to that for temperature, the map with hurricanes is of somewhat lower quality, and the twists and turns in the road are bigger, so it's not as easy to see that we're heading for the cliff. But it's still not wise to step hard on the gas pedal.

26

BARRIERS AND BEACHES SINCE 1938

The Long Island Express

On September 20, 2013, I rode Amtrak from New York to Boston. The next day was the seventy-fifth anniversary of the 1938 hurricane known variously as the Great New England Hurricane or the Long Island Express; to me it is just "1938." I was headed to an event commemorating it. The event was organized by the Blue Hill Observatory. A former National Weather Service station, the Blue Hill was decommissioned in the 1970s but has been maintained as an observatory since then by a nonprofit foundation. The people who run it are as proud of its long record of continuous observation as of its measurement of the 1938 storm's winds: 186-mile-per-hour gust and 121-mile-per-hour sustained for five minutes. Though these don't count for the purposes of estimating the storm's official intensity—made at the top of a hill, they aren't true surface winds—they are the strongest winds measured anywhere in the storm.

Like so many Atlantic hurricanes, 1938 formed in the Caribbean from an African easterly wave. The forecasters knew by the seventeenth of September, from ship observations radioed to the Weather Bureau, that it had become a hurricane.[126] Florida's residents were warned to expect the worst, but were spared when the storm turned north without making landfall. At that point, the storm was forecast to recurve to the east, out to sea, around the top of the Bermuda high.

Though the Weather Service's capabilities had advanced since the 1900 hurricane devastated Galveston with no warning at all, there were still no satellites, no aircraft reconnaissance, no computer models—no observations from anywhere but a few ships that might happen to cross the storm's path, and those had already been scrambling to get out of its way by the twentieth, the day whose seventy-fifth anniversary I was marking by my Amtrak ride.

The forecast of recurvature seems to have been based mainly on what we today would call climatology—that is, what storms in the past have done. It was wrong this time, spectacularly. This storm not only headed straight north, on a collision course with Long Island and New England behind it, but it did so with what, for a hurricane, is blinding speed. The storm raced parallel to the coast, covering the northward distance about as fast as one could do it in a car on Route 95, staying up all night: at least fifty miles per hour, maybe seventy at times.

This sprint gave the forecasters little time to see what was happening. As with Sandy, no warning of a hurricane was issued north of the Mason-Dixon Line. In 1938, though, that wasn't just a technicality. Whereas in 2012 only the hurricane label was missing from forecasts that otherwise spelled out the hazards clearly days ahead of time, in 1938 there was no real alert whatsoever before the center crossed the coastline. Although one junior forecaster, Charles Mitchell, apparently did produce a forecast on the twentieth that correctly predicted landfall, he did it just as a training exercise and put it in an envelope that wasn't opened until later. The actual forecasts issued for Long Island and New England contained no hint of what was coming.

The other effect of the 1938 storm's fast motion was to maintain its intensity.

Two things make Long Island and coastal New England, despite their exposure to the Atlantic, much safer from hurricanes than Florida, the Gulf Coast, or the Carolinas. The first is the tendency for jet stream westerlies to carry the storms offshore. The second is the cold water: sea surface temperature drops quickly with latitude; even in September, when the waters off Long Island are at their warmest, the maximum potential intensity a hurricane can sustain there is low.

But if the hurricane moves fast enough, it doesn't have to sustain its intensity for long. As it moves over the cooler waters, it takes time to adjust to its new environment. At highway speeds like those of the 1938 storm, a major hurricane can reach land before quite realizing how far from the tropics it has come. If a category four hurricane ever reaches New York, New England, or New Jersey, it will be a fast-mover like the Long Island Express was, one that just hasn't felt the weakening influence of its new environment yet. The 1938 storm made landfall in Long Island at category three, the only Atlantic storm in recorded history to reach these latitudes at that intensity.

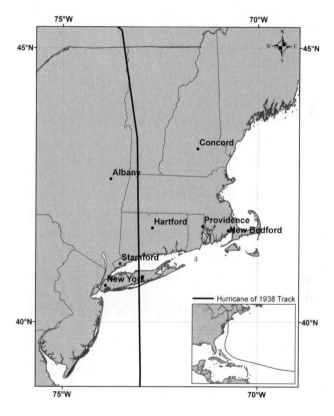

Map of the track of the Great New England Hurricane of 1938. *(Courtesy of Andrew Kruczkiewicz and Jerrod Lessel, track data from NOAA)*

The Amtrak Northeast Corridor line runs east along the coast, through Connecticut and Rhode Island, before turning inland around Providence.

If you ride it, sit on the right side of the train as you go from New York to Boston, the left coming back the other way. The shoreline here is deeply corrugated with estuaries, bays and inlets, remnants of rocky New England's ice age past. Glaciers carved channels well out into what is now the ocean, but wasn't ocean then. The ice sheet had locked up enough water to lower the sea by 125 meters and extend the coastline well out into today's Atlantic. These glacial scars remained when the ice melted and the sea rose and rivers carried sediment to fill in and smooth the landscape. The train runs along a line just north of most of the wider north–south inlets, minimizing the number of bridges while still hitting the major coastal towns: Stamford, Bridgeport, New Haven, New London, Westerly. Views of deciduous forest are regularly interrupted by rivers opening to the sea or by salt marsh, still undeveloped, cut by sinuous, meandering channels.

Nineteen thirty-eight's surge ran up hard into all those places. When a cyclone crosses this coast, the open, south-facing shores get the highest winds and the most powerful waves; but the highest surges, the worst floods, are in the bays and inlets, where the inrushing water becomes narrowly focused.

On Long Island, the storm made landfall as a category three hurricane. Images from the time show near-total destruction of shoreline communities. Houses were swept out to sea, buried in sand, or reduced to rubble. Similar images were taken all along the beachfront towns of New England's south-facing coast, in Connecticut, Rhode Island, and Massachusetts.

The storm weakened as it moved inland, but remained destructive. Though the best modern reconstructions still label the 1938 storm as a true hurricane at landfall, it was in the process of an extratropical transition; as it moved north into New England's interior, its peak wind speeds dropped, but the area of gale-force winds expanded. In the forests of northern New England, millions, perhaps tens of millions, of trees came down in the wind.

Nineteen Thirty-Eight also brought heavy rain. Coming on top of soils already soaked from a stretch of wet weather the previous week, the new downpours quickly ran off into the rivers, causing widespread inland flooding.

All told, the 1938 storm killed around six hundred people.[127] Many thousands of buildings were destroyed, many more damaged. Railroad service between New York and Boston was down for one or two weeks. The total

The Fox Point Hurricane Protection Barrier, Providence, Rhode Island. This photo predates the construction of the bridge that now carries Highway I-195 just next to the barrier. *(Courtesy of the U.S. Army Corps of Engineers)*

The Stamford Hurricane Protection Barrier, Stamford, Connecticut. *(Courtesy of the U.S. Army Corps of Engineers)*

economic damage was around three hundred million dollars, or four billion in today's dollars,[128] making it at least one of the most costly natural disasters to have occurred in history up to that date, if not the most.[129] The relief-and-recovery effort, carried out largely by the Works Progress Administration—President Roosevelt's job creation and public works agency was in its last years, as the Depression would shortly end with the arrival of World War II; FEMA would not be created for another forty years—and the Red Cross, lasted three months.

On Rhode Island's south-facing coast, east of where Point Judith juts out, the coastline drops back to the north into the wide expanse of Narragansett Bay. Several substantial islands partially fill the bay; the town of Newport is on Aquidneck Island, the largest one. Behind them, farther to the north, the bay narrows quickly as it reaches Providence Harbor. Two rivers empty into the harbor, converging at Fox Point. The Blackstone comes in from the north, passing just east of downtown; the Woonasquatucket comes in from the east, the point at which its channel begins to widen, defining the center of the city.

On the late afternoon of September 21, 1938, the storm surge flowed up-river and directly into downtown Providence. It arrived just as people were leaving work, around 5:00 p.m., flooding the streets suddenly. Cars floated; trolley cars, the public transport of the day, were swamped. City Hall, Exchange Place, the Biltmore hotel—the heart of the city was underwater.

Much of the worst damage and the most deaths in New England were in the coastal towns, where the winds were strongest and the waves the most powerful. But Providence wasn't the only town well back from the shore to flood. Stamford, Connecticut, and New Bedford, Massachusetts (neither one a state capital like Providence, but both substantial towns), also have estuaries that begin to open out in their downtowns. Just as in Providence, the surge funneled into both, putting the city centers under deep water.

Send in the Corps

Nineteen Thirty-Eight changed the perception among New Englanders that hurricanes were just for people in Florida to worry about—that "it can't happen here." It was also a turning point for the involvement of the federal government in coastal protection.

Up to the early twentieth century, the exposed coastlines of the United

States had been inhabited sparsely, home mostly to fishermen and a few rich owners of vacation homes. As automobiles became more common, populations (permanent residents and seasonal visitors) began to grow, building the constituencies to whom beach erosion—a fact of life anywhere land and sea meet in a strip of sand—mattered.

The New Jersey Shore was the first place in the country where beach tourism became a substantial industry. In the 1920s, the New Jersey state legislature commissioned studies of beach erosion, followed at the federal level by one commissioned by the National Research Council. Douglas Johnson, a geologist in what would become my department at Columbia University, was a contributor to the NRC study. A private nonprofit group, the American Shore and Beach Preservation Association, formed in 1926, would advocate for a federal program to protect the shore; and in 1930, Congress commissioned a new group within the Army Corps of Engineers, the Beach Erosion Board, to work more on the problem. The costs of the board's work would be shared equally between the states involved and the federal government.[130]

All this momentum led to legal action by 1936. A law passed then authorized the federal government to spend money on engineering projects for coastal protection, but not to maintain them once built.

Nineteen Thirty-Eight showed what a major hurricane could do to the shore. Not just houses, roads, and towns, but the coastline itself was reconfigured by the storm. Around a dozen new inlets were cut through the barrier islands on Long Island's South Shore; Shinnecock Inlet, in eastern Long Island, still exists today.

Then the war came. Protecting the coast from U-boats may have seemed more important than protecting it from hurricanes. A major hurricane did hammer the mid-Atlantic states in 1944, but it would take until the 1960s for the legacy of 1938 to be made, literally, concrete in the Northeast.

After the 1938 and 1944 hurricanes, new laws were passed strengthening the federal role further. Then, in the early 1950s, no fewer than six hurricanes hit the northeastern U.S. coast. In New England, Carol, in 1954, was particularly fierce.

Carol came up from the Caribbean sandwiched between a digging trough over the eastern United States and a high offshore. The flow pattern at 500 hectoPascals, in the days before landfall, resembled Sandy's quite

closely. There was no left turn, quite, but the trapping in the strong souther-
lies between the inland low and the offshore high kept Carol from recurving
out to sea. A winter-type shortwave low with a strong cold front behind it
probably also gave it new strength as it came up over the cold waters (again,
as with Sandy).

Though not quite as powerful as 1938, Carol caused a repeat of much
of the damage in New England. Again, the same coastal towns were hammered; again, the centers of Providence, Stamford, and New Bedford were
deeply inundated.

After Carol, and the other Northeast hurricanes of the early 1950s, another new law, passed in 1955, specifically instructed the Corps to consider
hurricane hazards. It specifically instructed that a wide range of measures,
from better forecasts to "breakwaters, seawalls, dikes, dams and other
structures," be considered.

Our Barriers

After Sandy, the media reported extensively on the storm surge barriers in
the Netherlands, as context for proposals for similar structures in New York
Harbor. But if New Yorkers want to see movable storm surge barriers, we
don't have to take an airplane. There are three nearby, in New England.
(Two of them on the Amtrak Northeast Corridor line, so you don't even
need a car to see them.) In the 1960s, the Army Corps of Engineers built
them to protect Stamford, Providence, and New Bedford, all cities whose
downtowns were submerged in 1938, and in 1954 by Carol. In September
2013, I went to see the Providence and Stamford barriers.

Each barrier works a little differently. Providence's Fox Point Barrier, underneath a bridge carrying highway traffic, has three scoop-shaped plates
that normally sit above the water but can swing down to close the channel.
Stamford (and New Bedford) has abutments (essentially partial dams, made
of rock) that run into the water from the natural shores on either side, nearly
connecting in the center but leaving gaps wide enough for ships to pass. Stamford's actual barrier is a plate that sits normally on the sea bottom between the
abutments and is pulled up into place by hydraulic motors to seal it.

Though they must have seemed large engineering projects at the time,
the New England barriers are small compared to those in the Netherlands.
The same reason the surge was so bad in Providence, Stamford, and New

Bedford—that is, the narrowing of the channels to the north, where they enter their respective downtowns, funneling the incoming water to a point—also meant that only a relatively narrow channel had to be gated. Barriers make obvious sense in these places, where high-density, valuable pieces of real estate can be protected by structures no bigger than a modest highway bridge.

For most of the coast, though, this doesn't work. The barrier islands and continental shores of New Jersey, New York, and New England are long and exposed. Protecting them with movable barriers is not an option.

Ash and Sand

Some of the "other structures" mentioned in the 1955 law that led to the construction of the New England barriers were described in great detail in a manual published a year earlier, in 1954, by the Corps' Beach Erosion Board. The manual outlined the science of beach erosion, and went on to provide specific guidance on what sorts of projects could be effective and how to build them.

Along with groins (barriers sticking out perpendicular to the beach, to trap sand flowing alongshore and keep it in one section of beach), seawalls (barriers along the shore), and other hard measures, the manual, entitled *Shore Protection Planning and Design*, described the potential benefits of adding sand to the beach. This practice was, and still is, known as beach "nourishment" (as if the sand were being eaten by the beach, instead of carried away by the currents). As well as adding sand right at the shoreline, the manual proposed building artificial dunes, mounds of sand set back from the shore. The report stated that:

> A belt of sand dunes will provide effective protection from upland property as long as the tops of the dunes remain above the limit of wave uprush. At those locations which have an adequate natural supply of sand and which are subject to inundation by storm tides and high seas, a belt of sand dunes may provide more effective protection, and at lower cost, then [*sic*] either a bulkhead, seawall or revetment.[131]

This "soft" engineering measure, however, had not reached its moment yet, but that would come less than a decade later.

Gary Szatkowski, meteorologist in charge of the New Jersey office of the National Weather Service during Sandy, put one historical reference in his "personal plea" to shore residents, as Sandy bore down on the New Jersey Shore in 2012: "If you are reluctant to evacuate, and you know someone who rode out the '62 storm on the barrier islands, ask them if they would do it again." The storm Szatkowski was talking about, the last truly major weather disaster for the New Jersey Shore until Sandy, had happened a half century earlier, in March 1962.

March is as far as the calendar gets from hurricane season; the "Ash Wednesday Storm" was a winter nor'easter, a pure extratropical cyclone. This storm would further strengthen the hand of the Army Corps in coastal protection engineering and would establish artificial dunes in particular as a prominent tool.

Huge and slow moving, the 1962 storm caused a surge that lasted for five tidal cycles; the storm's other historical nickname became the Five High Storm. Nearly the entire Eastern Seaboard was affected to some degree; New Jersey and Long Island were hit particularly hard. The descriptions and images from the event are Sandy-like: houses swept away by the sea, the Atlantic meeting lagoons across barrier islands, the shoreline itself substantially altered.[132]

A new law passed soon after the 1962 storm established that the federal government would pay 50 percent of the cost of coastal erosion-control projects—70 percent for public lands without private homes or businesses on them. The Beach Erosion Board was replaced in 1963 by the more substantial Coastal Engineering Research Center, first situated in Washington, DC, then moved twenty years later to Vicksburg, Mississippi (where it still is, though now consolidated into the broader U.S. Army Engineer Research and Development Center). Beach nourishment and artificial dunes became a large-scale enterprise.

Sand and Sandy

In 2009, a study done by a private engineering firm[133] for the City of Long Beach, New York, the next barrier island east of the Rockaways, gave a history of Army Corps studies that starts with one in 1965, motivated by the Ash Wednesday Storm.

A set of groins built along Long Beach Island in the 1940s had been effective at keeping sand on the beach. The study's executive summary noted that "the shoreline of the City of Long Beach has been generally accretional" (accumulating sand rather than losing it) since the groins were built. But, the summary went on, "beach elevation is not high enough to prevent flooding and coastal protection measures one [sic] warranted." In addition to new seawalls under the boardwalk, the study recommended "dry beach fill placed as a berm cap"—that is, artificial dunes. The Corps itself had made similar recommendations three years earlier, in 2006.

The dunes were never built in the City of Long Beach. As per the laws passed half a century earlier, the federal government would have paid much of the cost of the project, but the City Council voted against paying its share. The dunes would block the view of the sea.[134]

Long Beach Island was hammered by Sandy. The City of Long Beach, a town of thirty-three thousand, was estimated to suffer two hundred million dollars in damage. Houses were destroyed by fire as well as the surge; the boardwalk was obliterated. A video shot by *New York Times* reporters in the days after the storm shows the town in the same primitive state as most of the other hardest-hit communities.[135] The town had no power, water, or sewage; FEMA distributed bottled water amid the wreckage.

Lido Beach, Point Lookout, and Atlantic Beach are smaller towns that share Long Beach Island. Unlike their neighbor, the City of Long Beach, all had taken the Army Corps up on the offer of artificial dunes. These towns were reported to have done much better. Initial surveys by the Army Corps (perhaps not an entirely disinterested entity) found stark differences between the parts of the island with and without dunes. One CBS report, nearly a year after the storm, showed a resident standing by an intact beachfront home. The house, she was confident, had been saved by the second of two twenty-five-foot dunes, after the first one failed.[136]

Dunes were, and are, a bigger story still on the New Jersey Shore. Perhaps the most dramatic story was from Harvey Cedars, on New Jersey's Long Beach Island (not to be confused with New York's Long Beach Island, of the preceding few paragraphs).[137] Harvey Cedars had been devastated in 1962. The mayor before and during Sandy, Jonathan Oldham, was old enough to remember it, and had strongly supported an Army Corps project to nourish the beach and build up the dunes. But in order for the project to go ahead, all the owners of beachfront properties would have to sign easements allowing it. Most did, but a few owners refused. Oldham used the power of eminent domain to authorize the project over their objections, and the Army Corps finished the project in 2010.

One couple sued the town for their lost view, and was awarded $375,000 in court. After a higher court upheld the award, the town took the case to the state supreme court. While it was there, Sandy hit. The house of the suing holdouts, Harvey and Phyllis Karan, survived, as did most others in Harvey Cedars. The Karans—remarkably, in the view of many reading about it in the press, me included—didn't drop their suit. They, like other shore residents who had been holding out against signing easements to allow dune projects, were vilified in the media.

Perhaps their fiercest critic, and certainly the most prominent, was Governor Chris Christie. He stated his case in a town hall meeting on Long Beach Island on April 30, 2013—on YouTube, the video taken during the event is titled "Chris Christie Calls Bullshit on People That Don't Want Sand Dunes." Christie used that expletive to characterize the claim that the easements would be used to allow the building of unwanted new public facilities such as bathrooms or hot dog stands. Christie said the real reason those who opposed dunes, such as the Karans, did so was because the dunes would block their views, and because they hoped for financial compensation:

"It's about money, 'cause it almost always is. But let me tell you about the money we really need to be concerned about. This state sustained thirty billion dollars in damage after Hurricane Sandy. We had three hundred and sixty-five thousand homes destroyed or significantly damaged. But worse than that, we had the lifetime memories of families up and down the Jersey Shore washed away forever; they'll never be able to replace it. We are not going through that again so you can sit on the first floor rather than the second floor and see the ocean."

Sandy tipped the battle over the dunes in the state's favor. The Karans, their earlier award overturned by the state supreme court, finally settled, compensated one dollar for their lost view. Governor Christie signed an executive order directing the attorney general's office to "coordinate legal action to acquire the necessary easements to build dunes." Any holdouts now will come straight up against the power of the state government. The governor's goal appears to be a continuous barrier of artificial dunes along the entire New Jersey Shore.

Castles in the Sand

When a major storm changes the layout of the shoreline (taking some of the beach away, or adding some where it wasn't, or cutting inlets through barrier islands), it can seem dramatic, even shocking, to those who don't live near a beach or have the opportunity to observe any particular one regularly over many years. A beach is land, after all, and we expect land to be constant, solid; the sea is what we expect to be always in motion. But a beach, really, is a part of the land over which the sea retains some ownership and to which the sea imparts transience.

When a beach is rearranged by the ocean, it is not an anomaly. It is the norm. The storms that do the most substantial rearrangements come rarely, if we measure time by human life spans. But on geologic time, they are frequent. Weaker storms come frequently, even by human schedules, and waves and currents are constantly moving sand onto the beach, off it, and along its length, from one stretch to the next.

When a house is built on a beach, someone gains a stake in halting that transience. The sand becomes, literally, the ground on which they live. Keeping people's homes from washing away is a cause that's difficult to oppose—how would we feel if we were in their shoes? We can choose to build things to keep the beach in place, as the Army Corps does, but we have to understand that success will require a constant struggle against the shore's natural state of motion.

Beach nourishment, including dunes, looks environmentally conscious. It appears to be a co-opting of nature's own methods. But nature's beaches and dunes don't stay still. To make the artificial ones stay still requires constant maintenance. For a line of dunes along the New Jersey Shore to be worth the investment, it will have to be backed by a long-term commitment to keep replenishing them.

A barrier island (such as the Rockaways, Fire Island, or almost the entire New Jersey Shore) is a strip of land that is nothing but beach. It is an inherently transient piece of geography. It is just a glorified dune.

At the peak of the ice age, twenty thousand years ago, when the giant Laurentide Ice Sheet had a 125-meter-thick layer of ocean locked up and sea level was that much lower,[138] our barrier islands were parts of beaches, connected to the mainland. Then the climate warmed, ice melted, sea level rose quickly, and the coast retreated. As sea level rise slowed, around six thousand years ago, the barrier islands formed out of former beach sand left offshore—perhaps having been high dunes at the foot of the ice age beach, then separated from the mainland by the rising sea.

Barrier islands have never been stable. If one of them stays in one place for a while, it is just because the currents happen to be adding sand and taking it away at the same rate. In many cases, their days would be numbered even without climate change. Big storms move sand from their ocean shores to their landward sides. This can happen either with true breaches (where new channels are cut through, allowing sand to flow from the seaward shore to the landward bay) or with "overwash fans" (where storm surge cuts through dunes overland, blasting a ray of sand from the front to the back side of the island). Either way, this sand never goes back. Over time, the island migrates toward the continent, eventually to be reabsorbed.

Ocean City, Maryland, is on a barrier island. Fully developed, the town has maintained the beach by nourishment. Across Ocean City inlet to the south, an otherwise identical barrier island—they were one at some point, before a breach—is undeveloped. The southern island sits around five hundred meters closer to the shore than Ocean City; the disconnect is dramatic in aerial images. Cheryl Hapke, a geologist with the U.S. Geological Service in St. Petersburg, Florida, told me that this is a classic image for scientists who study beach erosion. It demonstrates with graphic clarity that development on barrier islands requires continual resistance against nature.

Hapke has been studying Fire Island, a New York barrier island east of Long Beach, for over a decade. On October 29, as Sandy bore down and residents were evacuating, she flew in and began making measurements to document the state of the beach. After spending the night of the storm in an empty hotel, she was out the next morning to record the changes to the island, before the bulldozers came in to start fixing the damage. Her team used lidar (an instrument that uses pulsed laser light to measure distances very precisely), aerial photos, and on-the-ground GPS surveys to measure precisely the elevation of the beach and also the depth of the sea bottom. Before and after images show how much sand has been moved from where to where.

Hapke's results[139] show that Sandy scooped away much of the beach on Fire Island. In some places, enough sand was removed to lower the elevation by one to three meters (three- to ten-foot drops in the height of the beach).

Much of that sand was washed out to sea; that sand is likely to be eventually recycled, rebuilding the beach, perhaps over the next year or two. Some recovery had taken place already by spring. Some sand, however, was blown back in overwash fans, or carried through a new channel made when the storm breached the island. That sand will not come back.

Google Earth image showing Ocean City, Maryland. The southern half of the barrier island has migrated landward since the development of the northern half, which has been held in place by beach nourishment. (© 2014 Google, TerraMetrics)

On average, Sandy moved Fire Island's peak elevation point about sixty feet toward the Long Island "mainland." The houses or trees or roads on the island didn't move, unless they were directly uprooted by the flooding, but the sand carried from the ocean to the bay displaced the island as a whole. This is the normal way barrier islands move over time. So with the understanding that no individual piece of real estate moved if the water didn't actually wash it away, it is fair to say that Sandy moved Fire Island sixty feet.

Hapke estimates, extrapolating from what she measured on Fire Island, that about thirty-seven million cubic yards of sand was removed from the beaches of New York and New Jersey by Sandy.[140] While much of that would eventually come back, it wouldn't necessarily go back where or when it's desired for protection of the structures by the shore. At least some of it will have to be replaced to keep the barrier islands, and the mainland beaches, where they are. That number gives a sense of the magnitude of the task. As sea level rises, much more rapidly now than at any time in the last several thousand years, the task will become much more challenging still.

27

IN HARM'S WAY

Although we can't yet detect a clear long-term upward trend in hurricane activity, either in the Atlantic or globally, there is an enormous trend in the *damage* caused by hurricanes.

Economic losses due to hurricanes (in the United States, in particular) have skyrocketed in the last several decades. The trends are almost certainly similar elsewhere, but are documented better in the United States. The reason has little or nothing to do with the hurricanes themselves. It is simply that more and more people are living, working, and building expensive stuff along hurricane-prone coasts. The rapid increases in hurricane damage in the recent past and at present (and likely in the near future as well) have been driven by increasing coastal development.

Hurricane-induced losses, in inflation-adjusted dollars, were much greater in the 1950s, '60s, '70s, and '80s than they had been up to that point. Then the season of 1992, when Hurricane Andrew struck Florida, saw an economic damage total far greater than that in any season previously. That record was then beaten by the 2004 season, which in turn was entirely blown away in 2005, which featured not only Katrina, but also Rita and Wilma—either of which would have been recognized as a major catastrophe in its own right had it not come in Katrina's shadow.

But these recent hurricanes were not demonstrably worse than those in the past; they just hit changed coastlines. The clear long-term trend in damage can be explained by increasing trends in population and development in hurricane-prone areas.[141] Globally, not just hurricane disasters but *all* natural disasters, defined consistently, have been increasing rapidly in frequency since the early twentieth century.[142] If earthquakes and tsunamis have been causing increasing harm to people just as hurricanes have, climate change can't have been the primary reason yet (though it is going to become a greater and greater factor as global warming proceeds and sea level rises). Even going back more than a century, we can see the consequences of development. When Galveston, Texas, was flattened without warning by the 1900 hurricane, it was a new city, growing rapidly as it tried to compete with Houston for dominance of the local economy.[143] Had the hurricane occurred fifty years earlier, it would not have killed the many thousands it did. Ninety-two years later, most of Andrew's damage was in Miami-Dade County, Florida, much of it in areas that would have been sparsely inhabited a few decades earlier. The city of Miami was spared the worst of Andrew; the last truly major storm to make a direct assault on Miami wrecked the city in 1926. If the same storm happened today, it has been estimated that it could cause twenty times the economic damage.[144]

Sandy caused losses currently estimated to be in the range of fifty to sixty-five billion dollars. A good fraction of that occurred in New York City.

The last truly comparable storm to hit the city was the 1821 hurricane. Its peak surge may even have exceeded Sandy's, but it came at low tide. While the total storm tide may have been slightly below Sandy's, it was certainly enough to cause substantial flooding in Lower Manhattan. As a disaster, in human terms, the 1821 storm clearly wasn't an event anywhere close to Sandy's magnitude. It is difficult to find much about it in histories of the city. The difference in the results of the two storms was unquestionably more due to the difference between the two cities, New York past and present, than to that between the two storms. The accounts of 1821 at the time describe considerable damage, but there is no comparison between the city then and the city now. There were no subways to flood, no electric lights to go dark. The population of the city was a little over a hundred and fifty thousand in 1821,[145] compared to eight million today. The wetlands of Staten Island were still wild swamps, no housing projects sat on the Rockaways, and there were no boardwalks or weekend homes along the Jersey Shore.

Disasters Waiting to Happen

Population growth and economic expansion are deep and powerful forces, acting over the long sweep of history. Coasts attract much of the growth. Transport of goods by ship is essential to commerce, historically and still today, and ports tend to grow cities around them. People like living near the water. Particularly in crowded, dense urban environments, the shore provides open space that one can see, even if one doesn't inhabit it, and a sense of contact with the larger natural world. Those who can manage to live by the beach or who have a second home near one value it tremendously— perhaps now more than ever.

New York City, in particular, has been reopening waterfront for public use aggressively in the last couple of decades. The Hudson riverfront in Manhattan, for example, is now lined with bike paths and parks for almost the entire length of the island. Having grown up here in the 1970s, when the Hudson was mostly inaccessible, I appreciate this dramatic change greatly.

The threats posed by hurricanes and other rare natural disasters are, most of the time, weak compared to the forces that propel coastal development. Still, with hindsight, after each disaster, it seems as though it should have been possible to develop in such a way as to avoid the worst vulnerabilities, particularly when the risk was well known long before the disaster occurred.

After Katrina, there was great outrage at the failure of the levees. The Army Corps of Engineers had, for many years, taken the construction and maintenance of those levees, the ones designed to protect from storm surges, much less seriously than the ones along the Mississippi River, designed to protect the city from floods due to rain upstream. Why was that, and why wasn't there any public complaint about it before the disaster? With a city largely below sea level on a very hurricane-prone coast, the threat was not hard to see. Scientists and others who understood the situation were concerned long before 2005. Max Mayfield, the director of NHC at the time, said after the fact,[146] "The 33 years that I've been at the hurricane center we have always been saying—the directors before me and I have always said—that the greatest potential for the nightmare scenarios, in the Gulf of Mexico anyway, is that New Orleans and southeast Louisiana area."

In 2005, the *New York Times* published an article[147] that began, "The national commission that studied the terror attacks of Sept. 11 concluded that

the lapses in preventing or responding to the attacks stemmed from a collective failure to imagine that such a catastrophe could happen." It went
on: "Emergency officials and meteorologists fear a similar failure on Long
Island about major hurricanes."

The article considered the potential harm to the region from a category
four hurricane. After cataloging the extensive destruction that would occur
in Long Island, the reporter, John Rather, wrote, "That would still pale in
comparison to the damage in New York City, where the Rockaway peninsula, parts of Wall Street, many subway stations and all of John F. Kennedy
International Airport would be under water."

Articles like this, about hypothetical future disasters, tend not to have
a great impact. This article in particular, though, didn't have any chance at
all. It was published on August 28, 2005; Hurricane Katrina made landfall
in Louisiana the next day, on the morning of August 29. Warning of New
York's vulnerability to a hypothetical hurricane instantly became overwhelmingly overshadowed by the very real disaster in New Orleans and
along the Gulf Coast. As we all now understand, that didn't make the warning about New York any less accurate or less prescient.

Water Flowing Underground, Once in a Lifetime (or Less)

New York City is not New Orleans. It's all above sea level, and the risk of an
intense hurricane happening there is much less. Still, the risk was known
to be there. The city had experienced a serious hurricane surge in 1821.
The 1938 hurricane was a close call for the city. There had been other close
calls, storms that either missed the city by a little or weakened just enough
before reaching us to become minor instead of major events. Sandy was an
unusual storm in its details, but clearly there was evidence that a storm of
this magnitude could reach us. Why was our infrastructure so unprepared
for it?

It's useful to focus on a specific example. As a native New Yorker who
didn't get a driver's license until I was twenty, I find the subways and commuter trains particularly compelling.

The subway system was built in the early twentieth century. The design
of the system shows no evidence that those who planned and built it thought
seriously about the possibility that it could flood. As we all know now, the
network of connected tunnels opening to low-lying stairways and air vents

constitutes a perfect system not just for transportation, but also for unintentional subterranean water storage. My generation inherited a vulnerable system. But that vulnerability wasn't newly exposed by Sandy. It was apparent for at least twenty years before.

In 1990, FEMA and the Army Corps conducted a study to determine the feasibility of evacuating residents from New York, New Jersey, and Connecticut in the case of a major hurricane. This study was novel in its use of computer simulation.

Similar studies had been done previously but had relied on historical records to assess how high a storm surge could occur. By their nature, such studies couldn't conceive of storms worse than any that had actually hit the city in the known past. For the new study, instead, NHC ran storm surge simulations with the SLOSH model, assuming a range of hypothetical storms. These had different tracks, intensities, and sizes, including combinations that hadn't occurred but couldn't be ruled out for the future.

The results were startling. NHC simulations showed that because of the shape of New York Harbor, a thirty-foot surge could not be ruled out. This was much, much larger than anything that had ever happened. Just ten feet would be enough to cause major flooding of the subways and commuter lines. These results shocked the deputy state director of the New Jersey Office of Emergency Management, and he asked FEMA and the Army Corps for more detailed studies of how a major surge would affect the area's transportation systems.

In December 1992, as these studies were ongoing, a major nor'easter (a winter storm) caused a storm surge large enough to flood parts of both the subway system and the PATH system that connects the city to Hoboken, New Jersey, briefly paralyzing much of the city. Though still minor compared to the worst that the NHC simulations had envisioned, this was enough to demonstrate that the vulnerability of the system was real.

In 1995 a report was published that outlined the vulnerabilities.[148] It featured photographs of some key low-lying subway and automobile tunnel entrances, with lines drawn on top of them showing the potential water heights that would be caused by hurricanes of different categories. The World Trade Center station (later destroyed in 9/11) would be submerged in a category three storm; the Hoboken PATH terminal, Manhattan's Holland Tunnel entrance, and the South Ferry subway station would in a category two, while the Manhattan entrance to the Brooklyn–Battery Tunnel would flood in just a category one. At that time, the Saffir-Simpson Hurricane Intensity Scale still included storm surge as one of the criteria for classification;[149] though Sandy's winds would only barely reach category one, its surge would have put it in category three.

Extensive tables documented the potential for inundation of all other major low-lying tunnels, rail lines, and stations in the tristate area. The report pointed out that because there are so many connections between tunnels, even parts of the rail system that didn't appear vulnerable at first glance were, because "interconnections at various levels ultimately combine all of the individual systems into one network." A statement in bold pointed out that "Nearly every rail tunnel system has significant points of entry below 10 feet NGVD," meaning a ten-foot surge would be enough to get into it.

The report made two sorts of recommendations. The primary purpose was to describe the steps that would be needed to evacuate people and save lives in the event of a hurricane. A key conclusion was that systems would have to be closed and evacuated prior to the start of the flooding. To allow effective evacuations, it would be best if a plan were implemented "whereby government and private business is closed prior to the beginning of the work day."

Together with all the recommendations focused on evacuation and other operational considerations during an event, there was a brief (one-paragraph) section entitled "Physical Protection from Surge." It read, in its entirety:

> Although the relatively short storm tide record for the Metro New York area is not sufficient to establish a surge height vs. frequency record with complete confidence, there is no doubt that moderate flood proofing measures in strategic locations would yield substantial dividends in terms of protecting transportation facilities and the public. A coordinated effort between agencies to flood proof vulnerable tunnel openings and raise roadways to a reasonable level could provide valuable insurance against shallow flooding. A project of this nature was recently undertaken along a portion of the Meadowbrook Parkway in Nassau County. Protection from coastal storm surge should be a consideration in all capital programs planning.

This report would not be the last to focus on these issues. In 2001 Columbia University's Earth Institute produced a report on the impacts of climate change (something not considered yet in the 1995 report) on the city and the region.[150] This report was written by a group of scientists led by my colleague Cynthia Rosenzweig, a scientist jointly appointed at Columbia and NASA's Goddard Institute for Space Studies, and William Solecki, then of Montclair State University. The lead author of the chapter on "Infrastructure" was my colleague Klaus Jacob of Columbia's Lamont-Doherty Earth Observatory. Sea level rise figured prominently, its impact on transportation systems in particular. Citing the 1995 study, the 2001 report explained that with a few feet of sea level rise, floods such as those envisioned in the 1995 report would occur much more often. Weaker storms, the kinds that occur more frequently, would be enough to cause them.

Further reports would follow. New York City and New York State became active consumers of climate information. Mayor Bloomberg, elected in 2000, would prove one of the most proactive elected leaders on environmental issues of all kinds at any level of government in the country, even forming a new city agency, PlaNYC, to deal with them. Climate change would rank highly on the city's agenda (along with bike lanes, smoking bans, and Midtown pedestrian zones).

Hurricane risk was on the radar, too, particularly after Katrina, in 2005, raised awareness nationwide. In 2007 the city's Office of Emergency Management cosponsored a design competition in which architects were asked to submit designs for long-term temporary housing that could be used by people displaced in a disaster. The trailers used by FEMA for Katrina refugees would, presumably, take up too much space to be practical in the densely populated New York metropolitan area. The competition asked the question "What if a Category 3 hurricane leveled an entire city neighborhood and left nearly 40,000 families homeless?"

Rosenzweig, Solecki, and colleagues produced more reports, under the aegis of new groups that sprang up to assess the impacts of climate change on cities: the New York City Panel on Climate Change, sponsored by the city; and the Urban Climate Change Research Network, sponsored by the United Nations and the World Bank. A 2011 report, sponsored by the New York State Energy Research and Development Authority, a state agency, featured the most comprehensive assessment to date of the potential for flooding of the subways. A chapter authored by Klaus Jacob and George Deodatis (also of Columbia, my colleague in the Engineering School) showed exquisitely detailed calculations that they and a group of their students had done, based on assumptions of storm surge from a moderate hurricane—technically, they assumed the historical hundred-year flood—on top of either two or four feet of sea level rise beyond present levels.[151] They computed which tunnels would fill, how long it would take to pump the water out (at least five days), and the time needed to return the system to nearly full capacity (over three weeks). They even estimated the total economic loss to the greater metropolitan area. Their estimates (between forty-eight and sixty-eight billion dollars) would prove eerily accurate.

Those calculations were tested much sooner than the authors expected. Though they had considered the flood due to a hundred-year storm tide on top of future sea level rise, a five-hundred-year storm tide would reach the same water levels without any new sea level rise, according to the FEMA maps in place during that time. Sandy was approximately a five-hundred-year storm tide event, according to those maps.[152] The projections proved highly accurate, almost down to the fine details. Jacob became a media celebrity, and has gone on to argue articulately and passionately for more serious action to prepare New York and other coastal cities for sea level rise.

The Jacob-Deodatis study devoted more words to long-term protective measures than the 1995 study had. The authors cited research showing that "there is an approximate 4-to-1 benefit-to-cost ratio of investing in protective measures to keep losses from disasters low," and estimated that this meant hundreds of millions of dollars per year should be spent on such measures immediately, rising to billions per year later. They were not specific about which measures were best, but suggested a range: raising critical infrastructure higher aboveground or relocating it to higher ground; building seawalls or levees to keep water out and pumping facilities to remove it; and "designing innovative gates at subway-, rail-, and road-tunnel entrances."

Since the Jacob-Deodatis report was written just a year before Sandy, it would have been unreasonable to expect the recommendations to be implemented by late 2012, even had they been adopted as high priorities immediately. But the essential message had been there in the 1995 report: a big storm could flood the transportation systems, with or without sea level rise.

The 1992 nor'easter had demonstrated it clearly, and the NHC calculations had shown that a hurricane could do much worse. Some of the recommendations of the 1995 report were clearly understood and adopted by the time of Sandy, and some weren't.

The need to shut down the transportation system ahead of time to get people out was understood and taken to heart. This was done for the first time in Irene and the second in Sandy. There can be little question that it saved many lives; it is terrifying to imagine what would have happened if people had been in the tunnels, either on trains or in stations, when these flooded. The roots of this decision go back to the 1995 report.

The recommendation to take proactive engineering measures to protect the system against flooding, on the other hand, was not followed in a serious way. The last-ditch efforts of the MTA to hammer plywood and heap sandbags over station entrances and air vents, heroic as they were, serve to illustrate that no more permanent defenses were in place. The construction of the new South Ferry station, opened just before Sandy, makes clear that this lack of storm defense was due not just to the difficulty of retrofitting a hundred-year-old system. Even when building a new facility, the MTA didn't take the risk of storm surge seriously. If they had, they wouldn't have built a new, five-hundred-million-dollar station with entrances in the hundred-year flood zone and no measures in place to seal them.

I don't write this to single out the MTA for criticism. Much of our national infrastructure is no more resilient than the New York City subways against the risks of natural disaster in our present climate, let alone the warmer ones to come. New York City under Bloomberg has actually been more proactive than almost any other local government about taking climate risks seriously. PlaNYC and the series of reports the city and state commissioned by Rosenzweig, Solecki, and their colleagues in the New York City Panel on Climate Change are models for the nation.

The MTA is not actually run by the city; it's a state agency. It is chronically underfunded by the New York State legislature, many of whose legislators come from upstate and don't see the city's subways as their top priority. As the agency struggles to maintain an old system while keeping fares affordable, it's hard to blame it for having failed to prepare adequately for either rare disasters or climate change. And it would be difficult to find examples elsewhere in the United States where essential public infrastructure on a scale comparable to the New York City subway system has been much better prepared for either.

Rather than being a particular failure of state or local government here in New York, the difference in the degree of adoption of short-term disaster management strategies versus long-term preventive measures in the subways illustrates a much broader and more profound problem. Individually and collectively, human beings are much better at dealing with threats that are in front of our noses than with those that are not, even when we know at a conscious level that the less immediate threats are real. This preference for the immediate appears to be deeply rooted in human psychology. It is very, very difficult for us to spend money or effort preparing for events that are outside our experience, even if we believe that they have a reasonable probability of occurring and that they could have disastrous consequences. Once something has happened to us, though, we become much better at preparing for the next time.

The evacuations and transportation shutdowns in Irene and Sandy seem at first to be counterexamples to the notion that we don't know how to react to something until it has actually happened to us at least once. These shutdowns were innovative measures taken, in fact, based on scientific understanding and prediction rather than experience. The system experienced some flooding in 1992, but not enough to kill people. Starting with the 1995 report, science was used to make projections about what could happen in worse storms. When such storms came, Bloomberg was able to give the right orders, and the MTA was able to execute them, without ever having been through a storm of the magnitude that Irene could have been and that Sandy was. But these successes didn't require major investments or actions far ahead of time; nothing had to be built, no capital projects undertaken. They required studies and planning, and those should not be undervalued. But neither are they in the same category as preventive measures to make the infrastructure itself more resilient.

Risk and Availability

The difficulty we have preparing for events outside our experience makes us inherently vulnerable, as a species, to rare events of all kinds. A five-hundred-year storm in any given place, by definition, is probably going to be outside the experience of everyone living in that place. It will probably be outside their great-grandparents' experience, too, and thus well beyond any meaningful collective memory. In our political struggles over spending public funds, it is extremely difficult for expensive infrastructure projects safeguarding against such storms (at least if one has not happened recently) to compete with schools, hospitals, the military, and tax cuts.

This is true even when preparing for such events is clearly rational. Economists (and insurance companies, not known for romantic environmentalism) define risk as the probability that something will happen multiplied by what it would cost if it did. Imagine we live in a city where we face two risks. The first is a rare one—let's say an eruption from a volcano just outside town with a one-in-five-hundred, or 0.2 percent, chance of happening in any given year. The second is a frequent one, which happens about once every other year, or 50 percent of the time—let's call it a snowstorm. But the cost of the volcanic eruption would be a billion dollars, while plowing the streets after a snowstorm costs a million. Given that 0.2 percent of a billion is two million dollars and 50 percent of a million is five hundred thousand dollars, the volcano is the greater risk, by a factor of four.

Now imagine that there has been no eruption in the last hundred years, and that the city was built a hundred years ago, so no one living there has experienced a volcano. The volcano has been quiet as long as anyone can remember. It has hiking trails and campgrounds on its flanks. Scientific research, on the other hand, has shown clearly that eruptions have occurred in the past. The average return period of five hundred years can be clearly established from deposits of volcanic rock in the area, which can be dated accurately. A new eruption would not be predictable more than a few days in advance. Evacuations are feasible to save lives, but the city itself would be devastated.

Now imagine that the city has a fixed budget to divide between snowplows and long-term measures that could reduce the harm from a volcanic eruption. How do you think the budget will be allotted?

Once something happens, though, the story changes. The hypothetical becomes real. For a little while, people gain the ability to think in the long term.

Studies of insurance show that many people tend to buy policies immediately after a disaster occurs, even though premiums tend to rise at those times. However, owners of new flood insurance policies (required in order to obtain federally backed mortgages in some areas) tend to let them lapse after a few years if no flood occurs.[153] Both these behaviors are irrational. They would make sense if the odds of a flood this year were to increase if there was one last year, and decrease if there was not, but that is not the case. The odds of a flood this year are the same regardless of what happened last year. One explanation for this is a feature of human psychology known as availability bias: people tend to estimate the probability of an event based on their ability to recall instances of similar events.[154]

Sometimes the window in which the memory of a disaster is "available" stays open long enough for genuine investments to take place. The 1953 flood led to the construction of barriers in the Netherlands. It also institutionalized seriousness about modern flood protection there that has continued to evolve, leading to the Room for the River and other projects that take a less confrontational stance against the water. The Thames Barrier, designed to protect London, was built in response to the same 1953 flood that sparked the Delta Plan. (That flood killed three hundred people in the United Kingdom. Though much smaller than the Netherlands' toll, the number was large enough to induce action. The Thames Barrier was finished in 1982, a few years before the Oosterschelde.) The hurricanes of 1938, 1944, and then Carol and the others in the 1950s led to construction of the smaller barriers in New England. All those, plus the 1962 Ash Wednesday storm, led to the systematic federal effort to build sand dunes and nourish beaches along the United States' Eastern Seaboard. Hurricanes Betsy and Katrina led to the first- and now second-generation hurricane protection systems for New Orleans.

One could debate whether any of these solutions was the right one. That is a question of values. Whether it makes sense to commit to pouring sand on a beach in perpetuity to protect homes on a naturally transient barrier island, for example, comes down to how one views the relationship of human beings to nature. But at least something was done. The silver lining in the clouds of all these historical storms was the strengthening of defenses against the next one.

They weren't strengthened enough, clearly, or Sandy wouldn't have done as much harm as it did. And rather than consider retreat, coastal development continued to put more people, buildings, and infrastructure in harm's way. But what defenses we do have are the legacy of past storms.

Sandy will leave a similar legacy. New defenses will be built. New measures will be taken to make New York and New Jersey more resilient. What should these be? How should our understanding of the relationship of human beings to nature change after Sandy?

The challenge of thinking long term is exacerbated by climate change. The extent to which global warming influenced Sandy is an important scientific question, but not the most important one for practical purposes.

Even if warming played no role in Sandy beyond adding a few inches to the water levels via sea level rise, it should still be a major factor in our planning for the future. Warming is accelerating, the sea will rise farther, and the risk of a truly major hurricane, though still small in any given year, is quite likely increasing as well.

And warming will bring a range of other hazards. Some of them are much more certain than changes in hurricanes, and probably more damaging, if less spectacular. Having experienced many a summer heat wave in New York City, I'm truly frightened at the notion that what we now call a heat wave will, in a few decades, be seen as a relatively cool period, and that the heat waves of the late twenty-first century will be well beyond anything ever experienced in our history—predictions that are almost certainties at this point.

Our normal pattern of responding to long-term disaster risk won't work for global warming any better than it worked to prepare our infrastructure for Sandy, and the consequences of ignoring the risk in this case will be more serious still. Normally we don't take seriously warnings of any particular disaster until it happens; then we gear up for next time. Global warming builds up slowly, but it will bring us outside the realm of anyone's experience, permanently. By midcentury, for example, the coolest summers in many places are very likely to be warmer than the warmest summers that have occurred in the historical past. Our ability to prepare for such a future, one for which we have no historical precedent, will be severely hindered if we cannot overcome our availability bias.

Yet taking the approach of waiting until the event happens and then reacting to it will be even worse with regard to global warming than to other types of hazards. Global warming is not a single event, but a long-term change. It will be permanent, for all practical purposes. Once emitted into the atmosphere, carbon dioxide stays in the climate system for hundreds of years, perhaps thousands. If we reduce fossil fuel use now, when much of the warming we will experience is still in the future, we can reduce the extent of the warming that will eventually occur. But by the time the impacts of human-induced global warming are severe enough to convince every last doubter that something needs to be done, much more warming will be already locked in. Buying insurance after the flood will not work.

28

WHAT HAPPENS NOW?

So where do we go from here? How has this singular event changed how we—in New York City, in the northeastern United States, in the nation as a whole, in the world—see the weather, the climate, the risk of living by the sea? What will be different now?

It's useful to categorize the consequences of Sandy by the different time scales of response to a disaster. There are the short-term actions once we know a storm is coming: forecasting and warning; emergency preparation, management, and immediate response; and then relief. Then there are the long-term actions we can take to reduce the risk from storms in the future, beyond the horizon of today's weather forecast.

The Short Term

Forecasts and warnings are never perfect. As good as they were in Sandy, they could have been still a little better. It would have been better had NOAA's procedures not drawn such a sharp line between extratropical and tropical storms, so that hurricane warnings (the most effective, attention-grabbing public statements available) could have been issued for New York and New Jersey. It would also have been better if the storm surge forecasts had reached the six- to eleven-foot mark a day or two sooner than they did. If these things had happened, perhaps New York City might have grasped the full gravity of the situation a little earlier than it did. More people might have heeded evacuation orders, and perhaps nursing homes and other vulnerable facilities that weren't evacuated would have been.

For New York City, perhaps the most damaging imperfections in the weather forecast were those in the forecasts not from Sandy itself, but from Irene a year earlier. Irene's intensity at landfall was less than had been forecast, and its surge was at the very low end of what had been forecast. The authorities and the media had prepared the city for the high end. When the disaster didn't materialize, it made many people less sensitive the next time they heard similar warnings. As Sandy arrived, many didn't evacuate who should have because of that false sense of security.

Yet, given the scientific information, the warnings in Irene were appropriate. There is always some uncertainty, and it is right to prepare for the worst-case scenario. Communicating that inevitable uncertainty in a way that leads people to respond in the most effective way—so that the false alarms that do occur don't cause the next warnings to be ignored—is a challenge that remains for the future.

The emergency response and relief efforts could stand to be improved as well. The emergence of Occupy Sandy, and the critical role it and other volunteer groups played, can be viewed positively, as a testament to the community bonds and selfless values of those who gave so much of themselves to help others. But they also highlight gaps in the government response.

Yet I think it would be mistaken to focus on these short-term issues. One can always find things to criticize afterward, but in my view most of our short-term systems functioned remarkably well in Sandy, at least compared to any historical precedent.

The forecasts, overall, were extremely accurate. The local governments did most of the right things to prepare. And the relief effort was extremely focused and effective by comparison to that for the most recent similar event, Katrina in 2005.

One could object that the response to Katrina was an abject failure of government at all levels and shouldn't be used as a standard against which to measure anything, or that it is natural to expect New York City to function better than New Orleans. These are fair objections, but the response to Sandy also benefited from learning in the wake of Katrina—on the part of Barack Obama, who had seen the consequences of George W. Bush's failure in Katrina, but also within FEMA and in state and local agencies.

The evidence shows that our short-term systems for dealing with natural disasters are capable of continuous improvement. Forecasts keep getting better. The rules that prevented the issuing of hurricane warnings in Sandy have already been changed. If Sandy were to happen today, hurricane warnings would be issued for New York and New Jersey. The next time there is another similar event in this area, as long as we have governments in place (federal, state, and local) that are comparable in their competence to those we have now, it's reasonable to expect that the other aspects of emergency preparation, management, response, and relief will go still better than they did in Sandy.

These short-term issues worry me much less than those involved in preparing for the next Sandy, which are inherently long term.

The Long Term

Sandy was not just an extreme fluke, something that we can assume won't happen for another few hundred years.[155] But neither is it "the new normal"— something that is sure to happen again soon, and often from now on.

Almost certainly it's somewhere in between. We're very unlikely to see another Sandy this year, or next, or even in the next decade or two. We're not that much more vulnerable today than we were a few decades ago. But at the time of Sandy, we were always more vulnerable than we realized. And the pace of change is quickening.

Because of sea level rise, most of all, our risk of more Sandy-type disasters is increasing. The science of hurricanes and climate change is still young, and some of the features that made Sandy's surge so big (its enormous size, its hybrid character, the left turn and westward-tracking landfall) are among those whose connections to climate are least well understood. But because of sea level rise, we know that big coastal flooding events will become more frequent, almost regardless of what those connections are.

As far as the potential for flooding is concerned, every foot of sea level rise is equivalent to a substantial increase in storm intensity. Under the old Saffir-Simpson Hurricane Intensity Scale, when it still accounted for storm surge (before it was simplified to measure only the maximum wind speed), the step from category one to two, or the step from two to three, carried a three-foot increase in surge. In the next hundred years, we are reasonably likely to see a permanent three-foot increase in sea level, and even six feet is not at all out of the question. That's roughly equivalent to an increase of either one or two categories in hurricane intensity.

In other words, even if the storms don't change at all, the potential for flooding will be about equivalent, in the next hundred years or sooner, to what it would be if each storm were to ratchet one or maybe even two categories up the Saffir-Simpson scale. While pondering this, recall that Sandy, had it been classified as tropical, would have been a hurricane only at the low end of category one under the current scale (which measures only the maximum sustained wind speed). The massive flood was due to the wild-card factors: the size, the track, the bad luck of a landfall right near high tide. But increase the intensity, or raise the sea level, and those factors don't have to line up anymore in order to produce the same effect. A big tropical storm or a moderately strong nor'easter (the kinds of storms that cross our coastline quite often in the present climate, without our noticing them much) can easily produce a four- or five-foot surge. Add another four or five feet of sea level, and you have nine feet, which was the size of Sandy's surge. At high tide, one of these garden-variety storms would be enough to put the water in the same places Sandy did.

On the other hand, sea level rises slowly. We have time to prepare. If we adapt to it as it happens, then one foot of sea level rise in the future will not be equivalent to one foot of storm surge now, because we'll be better protected. We could put other defenses in place that would have the same effect as if we had raised our cities and towns along with the sea. Then, a four-foot surge in the future, like a four-foot surge today, will not be a disaster. That would be climate *adaptation*. In the language of climate policy, that word refers to any action taken to reduce the harm from warming.[156]

Even better, we could simultaneously do climate *mitigation*, the word used for preventing the warming from occurring in the first place. We, the population of Earth, could simultaneously reduce greenhouse gas emissions, by reducing our use of fossil fuels. If we were to reduce it enough, we could significantly slow the rate of global warming, and the rate of sea level rise. Some warming and some sea level rise are already locked in, because of the carbon we have already put into the atmosphere. But if we were to reach a serious international agreement to transform our energy systems to be more efficient, and more reliant on renewables such as solar and wind energy—or even nuclear, though that brings another set of risks—we could make a significant dent in the problem.

Will we do any of that, though?

Denial

It's hard to be optimistic about the chances for major reductions in greenhouse gas emissions in the near term (which is when we need them most).[157] The international negotiation process on climate, carried out since 1992 under the United Nations Framework Convention on Climate Change (UNFCCC), has to be judged largely a failure, if success means securing binding agreements to reduce global emissions that include all the key players. There have been no such agreements.

For too much of the history of the negotiation process, a big part of the problem has been the United States. Under President Clinton, the United States signed but did not ratify the Kyoto Protocol, the 1992 treaty that may have, in retrospect, marked the high point of the UNFCCC process to date. Since then, U.S. leadership on climate has been lacking. There are many possible reasons for this, some of them not unique to the United States. Reducing dependence on fossil fuels is a huge global challenge, and getting different countries to agree on how much each should do adds a great deal to the difficulty.

Too much of the American problem is the virulent strain of denialism: simple rejection of the science. This denialism is not justified by actual scientific uncertainty (though it claims to be). It appears to be motivated by dislike of the perceived political implications of the science. The book *Merchants of Doubt*, by Naomi Oreskes and Erik Conway, documents the history of the denialist movement in detail. Oreskes and Conway show quite clearly that the most outspoken and influential deniers of climate change are the same groups, and in some cases the same individuals, who have been the most outspoken and influential deniers of the science on nearly every major public health and environmental issue since the 1950s.

When American tobacco companies were faced with unequivocal scientific evidence that cigarette smoking caused lung cancer, they responded by creating their own pseudoscientific public relations campaign whose goal was to make it appear that the science was much more uncertain than it actually was. They hired scientists, sponsored research studies, created research journals, and placed advocates in committees charged with writing government-sponsored reports. But unlike the normal scientific process, in which scientists are charged with discovering the truth, the scientists paid by the tobacco companies were charged with undermining it. They were paid to create an appearance of uncertainty on questions that the actual science had already answered with more than enough certainty to justify regulation of cigarette smoking.

The tobacco companies eventually lost the battle over smoking, but they delayed the regulation of cigarettes by decades. Oreskes and Conway show that the same people, groups, and tactics were then used on second-hand cigarette smoke, on acid rain, and on the ozone hole, and are now being used on climate change. In each case, science advanced to the point where a serious risk to the public could be conclusively linked to some pollutant. But action was (and, in the case of climate, is) delayed by the denialists' intentional, and aggressively dishonest, sowing of doubt.

The motivation behind the denialist movement has an obvious self-interested component. Tobacco companies wanted to sell cigarettes, chemical companies wanted to sell the chlorofluorocarbons that were depleting ozone, and fossil fuel companies want to sell fossil fuels. But it isn't only that. Some of the most prominent denialists, particularly the PhD-bearing scientists, some of them with distinguished records in their earlier careers (though, almost without exception, in scientific fields not relevant to the environmental or public health issues in question), appear to have their own motivations, not purely self-interested. Oreskes and Conway draw the conclusion that many of these individuals are motivated by some combination of simple contrarianism—as one of my colleagues once said, "you can't be the smartest person in the room if you agree with everyone else"—and a deep, almost religious commitment to pure free-market economic principles.

Accepting that our industrial society produces emissions that are harming the planet seems to imply that government regulation is necessary to reduce that harm. If you believe that all government regulation is inherently evil, you may be led to the conclusion that the harm caused by those industrial emissions must be fictional, a lie created by a conspiratorial left-wing environmentalist movement.

What frustrates me about the climate denial movement is not the politics of those associated with it. It's the practice, conscious or unconscious, of basing one's understanding of the science on one's political views. We all have to strive to separate those things, or we can never have a productive debate on climate, or on any issue where science is relevant to the public sphere.

Scientists have political views, too, of course. But our training, and the peer-review system that all our work must endure, force us to justify our results on their merits. When the collective action of that scientific system results in a clear scientific consensus—and it has, on human-induced global warming—the result is a scientific fact. We all have to accept that and focus our political arguments on what to do in the face of this reality.

The fact that human greenhouse gas emissions are causing serious climate change poses an enormous challenge to the human species. It is a far greater challenge than any previous environmental problem has posed, both because of its global scope and because of the deep entrenchment of fossil fuels in nearly every aspect of our infrastructure, our economy, and our lives.

No one is pure. Climate scientists and environmental activists also drive cars, fly in airplanes, and use products made in fossil-fueled factories and transported in trucks. This is not a narrative with clear villains, victims, and heroes. We all are villains, all victims, and (maybe) all potential heroes. But doing anything requires collective action on a global scale. Many perspectives will be needed to find the best solutions.

Since the choices we will have to make involve value judgments, we need to be able to have serious discussions about those value judgments at the level of our national and international politics. Such discussions are seriously handicapped if major factions simply refuse to participate, instead declaring, against all the genuine scientific evidence, that the entire premise is false. But that is where we are on climate in the United States.

Climate change should not be a liberal issue. There are serious, intellectually honest conservative positions on climate change. If one believes that government intervention should be minimized, one can advocate for market-based strategies for emissions reduction. If one believes that the costs of some steps toward emissions reductions are too high compared to the benefits they will bring in reducing the long-term harm from warming, one can argue that. These are values questions, not science questions. But avoiding these questions entirely by simply denying the reality of the science is not an intellectually honest position.

The United States (and the world) needs a conservative movement that will accept the reality of climate change, and proceed from there to argue for the solutions that are most consistent with conservative values. Once we have that, we will be in a better position to take serious action on climate.

It will happen eventually. The climate keeps getting warmer, and reality can be denied for only so long. But it would be a lot better if it happened sooner rather than later.

Local

At the state and local level, there is much more positive movement on climate, and on environmental issues generally. In some states and cities, the denial-driven gridlock that has paralyzed our national politics is absent. Things can happen, and have happened.

On mitigation, states on the East and West Coasts have moved toward local emissions-reduction and -trading schemes. New York City under Mayor Bloomberg has taken significant steps to reduce carbon emissions.[158] These actions have only small impacts, in that they address only small fractions of global emissions. But they are at least symbolic, and represent some much-needed leadership. California has long led the rest of the nation on automobile efficiency regulations. The state makes its own laws, ones more aggressive than those elsewhere; the rest of the country, and the automobile manufacturers, eventually catch up. Those efficiency increases were originally driven by the health effects of ground-level air pollution, but have the additional benefit of reducing greenhouse gas emissions. Maybe states and cities can play a similar role in enacting other measures to reduce the rate of climate change. We can hope.

On adaptation, states and cities are the natural leaders. While global warming is truly global—a ton of carbon emitted anywhere has the same effect as a ton emitted anywhere else, warming the whole planet regardless of where it came from—all adaptation is local.

What measures will be taken to protect the New York metropolitan area from future Sandys is not just a question about adapting to climate change. Sandy made us aware of a risk that has always been there, since long before the atmosphere's carbon dioxide concentration began to climb. We would need to respond to that awakening even if the climate were unchanging.

Sandy could have happened without any climate change, and it would make sense to have stronger protections against coastal floods even if the climate were static. Even the most ardent climate change denier can't deny what happened in late October 2012, or the logic of being better prepared for the next superstorm.

At the same time, sea level rise (and perhaps warming-driven changes in storm activity) are happening now, and raising the stakes. The new momentum for coastal flood protection provides a new opportunity to prepare for these changes as well.

In New York City, the Bloomberg administration was well positioned to take this opportunity seriously. For the past twelve years, under PlaNYC and several related initiatives, including the New York City Panel on Climate Change, the city has assembled highly qualified and motivated teams of experts, both within the city government itself and in collaborating outside organizations, with the goal of making the city more sustainable in all ways, including increasing its resilience to climate-related hazards.

In December 2012, shortly after Sandy, Bloomberg ordered a Special Initiative for Rebuilding and Resiliency (SIRR), which he charged with producing a plan for "A Stronger, More Resilient New York." The SIRR report is clearly a response to Sandy. Its scope is narrower than the title implies, but the report is still remarkable both in its breadth and depth. A 438-page document, it will probably not be read from front to back by too many of the city's residents, but it is eminently readable, and profoundly comprehensive. I am not aware of a comparable document having ever been produced by a local government anywhere.

The report begins by describing Sandy itself and its impacts on the city. Then it outlines strategic goals of increased resiliency for each of eleven separate dimensions of New York's infrastructure,[159] and proposes measures to reach those goals.

The report proposes many, many things. There is no silver bullet; storm surge barriers across the harbor, in particular, are not proposed. Instead, it recommends a wide range of smaller, incremental measures. The report recognizes the reality of climate change: on the advice of the NPCC, it builds in the working assumption that sea level will rise 2.5 feet by 2050.

The report, motivated by Sandy, focuses largely on storm surge risk, but it also considers the increasing risks to each system from heat waves and heavy downpours as the climate warms. The report notes that "These projections have been subjected to rigorous peer review, and represent the best-available climate science for New York City. However, they are not yet officially recognized by the State or Federal governments because there is no formal mechanism for them to do so."

This is a somewhat charitable way of pointing out the leading role that New York City and some other localities are playing on climate—or, equivalently, the lagging at higher levels of government.

The Coastal Protection Planning in the SIRR report proposes nearly every possible measure in some part of the city. Under the plan, at "full build," a local storm surge barrier comparable to the Stamford, New Bedford, or Providence one, goes up in Newtown Creek, which runs between Brooklyn and Queens. Oakwood Beach in Staten Island and Flushing Meadows in Queens get tide gates. The Rockaways get dunes, and beaches also get nourished there, as well as in Coney Island and Staten Island. The plan has old-school hard measures such as bulkheads, groins, floodwalls, levees, offshore breakwaters, and revetments, but also restoration of natural features: "Wetlands, Living Shorelines and Reefs."

The "Buildings" section addresses the opportunities and the difficulties in making the city's homes and businesses more resilient to storm surge, particularly as the sea rises. Many buildings that were not considered to be in the hundred-year floodplain when they were built are in it now, and many more will be by 2050. The ideal remedy for storm surge risk is raising critical parts of buildings (living space, commercial space, utilities) above the levels where floodwaters can reach them. This is very difficult in many parts of the city. Many buildings are old, often packed close together, and can't easily be retrofit without being torn down. Commercial storefronts don't do very well if they are above the eyeball levels of pedestrians on the streets.

The "Insurance" section makes for some more difficult reading. Most of the buildings flooded in Sandy had no flood insurance; 20 percent of the residential buildings and an even lower percentage of businesses did. Many people who did have coverage found that they didn't have enough, or had technical difficulties getting claims paid.

In principle, everyone in the floodplain should have insurance. Since 1968, homeowners and small businesses have gotten flood insurance (if they do so at all) mainly through the federally sponsored National Flood Insurance Program (NFIP). Rates in many areas are subsidized, so that the premiums paid are less than what is needed to cover the expected long-term average claims. However, the Biggert-Waters Flood Insurance Reform Act (passed by Congress earlier in 2012, just before Sandy) reduced these subsidies, dramatically in some cases. Rates for many of those in the floodplain would have risen sharply due to Biggert-Waters alone, even if the underlying assessment of flood risk itself had remained static.

In fact, though, the floodplain is growing. The areas subject to the hundred- and five-hundred-year flood are defined in maps made by FEMA. FEMA increased those areas in 1983—a long time ago, but more recent than the construction of many of the buildings in the new, enlarged floodplains. Now, post-Sandy, FEMA is revising the flood maps again. The final versions have not been released yet, but the preliminary versions show significant new increases in risk—even without accounting for future sea level rise. This is excluded from the calculations going into the maps.[160] New maps drawn in the future, though, will presumably incorporate at least any sea level rise that has occurred up until then, if not estimates of additional future rise. So the insurance rate increases occurring now will not be the last, if the rates reflect the best scientific estimates of the actual risk.

This is a hard reality that will apply in many parts of the country, but one that Sandy has caused New York and New Jersey to face before most others. Their low-lying areas are inherently at high risk. This is even more so than was recognized in the past, and yet less so than it will be in the future. At the same time, with Biggert-Waters, the federal government decided that it would no longer subsidize high-risk areas to the extent it did until now.

The path of the Biggert-Waters Act shows the political difficulty of doing anything to slow (let alone reverse) the ever-increasing development of high-risk coastal areas. Biggert-Waters passed easily in one of the most politically divided and ineffective Congresses in U.S. history. Conservatives liked the act because it reduced a government subsidy; progressives liked it because it made the risks associated with climate change more apparent. Yet in 2014, as I was in the process of finishing this book, Congress moved to roll it back after the rate increases that followed drew intense protest from constituents in coastal states. The entire purpose of the act was to make flood insurance rates increase; yet the lawmakers who passed the act, apparently, did not grasp what that would really mean until it happened. On March 31, 2014, H.R. 3370, the Homeowner Flood Insurance Affordability Act of 2014, which reverses much of the Biggert-Waters Act, was signed into law by the president.

In any case, because flood insurance is mostly federal, it is not under local control. The proposals in the SIRR report reflect this; all involve appeals to or coordination with higher levels of government or with the private sector. The report proposes that the city "Support Federal efforts to address affordability issues . . ."; "Call on FEMA to develop mitigation credits . . ."; "Call on New York State to improve policyholder awareness"; and "Launch an engagement campaign targeting insurers."

Much of the report, in fact, illustrates that even the most forward-looking city government can do only so much on its own. Many of the most important measures require serious cooperation from other entities, both in state and federal government and in the private sector.

The "Utilities" section proposes that the city "Work with utilities, regulators . . ." and others to harden and raise key infrastructure (for example, substations such as the ones that shorted out in Lower Manhattan), and develop long-term plans for increasing resiliency as climate changes. The "Transportation" section, too, highlights that the city can't do that much on its own. Given the absolutely critical role of the subways to the city, and their highly visible failure in Sandy, one might have expected to see some substantive plans to harden the system. Instead, one short initiative is titled "Call on non-City agencies to implement strategies to address climate change threats." The non-city agencies covered by this single bullet point include the MTA, the state agency that runs the city's subways and buses; and the Port Authority, the multistate entity that runs the airports as well as the bridges and tunnels that connect the city to New Jersey.

The MTA's workers were true heroes in Sandy. But the subways shouldn't need heroism to survive or recover from a storm. They should be designed to be able to better withstand it in the first place, and that takes long-term planning. That planning now needs to account for climate change, especially sea level rise. It has not done so in the past, as the totaled South Ferry station demonstrates. The MTA, after years of coaxing from Klaus Jacob and others, now seems ready to take sea level rise seriously. The agency is building it into its capital planning for the first time.[161] But protecting the system against it properly will surely be very expensive.

Maybe the MTA, and the other city and state agencies that manage critical infrastructure, will now get some of the resources they need to address this challenge in a serious way. Like Mayor Bloomberg did in the city after Sandy, Governor Cuomo commissioned a group of experts to write a report on how the state should prepare for climate change and future storms.[162] That report informed a speech Governor Cuomo gave in early January 2014, with Vice President Biden in attendance, on how he would spend $17 billion in federal funds allocated to New York State for Sandy recovery. He proposed $5 billion to "fortify transit infrastructure."[163] One of the measures he specified was giant inflatable plugs to seal subway entrances.[164] Other ambitious measures include $257 million in flood protection for the city's airports and the establishment of a new university exclusively for emergency management.

Well-meaning and well-informed experts could debate whether these are the right measures, or whether the right amount of money is being directed to each. Regardless, it is good to see climate adaptation and coastal flood preparedness being given a new level of serious attention, at the state and local levels. What we need to see is both follow-through, as the memory of Sandy fades, and sustained support at the federal level as well.

AFTERWORD

A natural disaster has two moving parts. One is the extreme event that nature delivers. The other is the vulnerability of the human beings and their structures—physical, social, economic, and political—that lie in the event's path. The converse of vulnerability is "resiliency." The second page of the New York City SIRR report, after the title page, has nothing on it but a dictionary definition of the adjective *resilient*:

re·sil·ient [ri-zil-yuhnt] adj.
1. Able to bounce back after change or adversity.
2. Capable of preparing for, responding to, and recovering from difficult conditions.
Syn.: TOUGH
See also: New York City

Being capable of preparing for, responding to, and recovering from a disaster does not constitute three different things. All three are mostly the same thing. People, things, and institutions that are better prepared will respond and recover better.

Some of being better prepared is just being in better circumstances to begin with. When Sandy killed more people in Haiti than in Cuba, or proved more disastrous to those in nursing homes and housing projects than to most others in New York City, it was following a typical pattern. Comparisons of Sandy to the worst storms in poorer nations overseas (some with death tolls in the thousands, tens of thousands, or even hundreds of thousands) bring this point into sharper relief still.

The other part of being better prepared, though, is understanding what is coming. That is where science enters. But the science can be useful only if those in positions to make decisions take it seriously enough. We can look to our leaders for guidance, but in a democratic society, they look to us, too. All of us are in a position to make decisions. At the very least, we can vote.

How seriously do we take the science? The answer to that question depends on the time scale. We are capable of taking science quite seriously when it tells us that something is about to happen soon, at a specific time and place. We take it much less seriously when it tells us that something is likely to happen eventually, at some unspecified (or imprecisely specified) time in the future. But at this moment in history, we as a species need to learn to take long-term predictions of risk as seriously as we take short-term predictions of specific events.

Over the last hundred years, our scientific understanding of the atmosphere has evolved tremendously. For most of human history, weather prediction was based largely on experience, intuition, and folklore. Starting at the turn of the twentieth century, first in Scandinavia and then growing out from there, a small group of scientists had the vision that the rigorous disciplines of physics and mathematics could be brought to bear on the weather. This began a scientific revolution whose results were on clear display in the days leading up to Sandy's landfall.

When the models first envisioned Sandy's landfall eight days before it happened, media stories began to publicize that vision. Government officials began to take the event seriously as it drew closer. As the spaghetti of model tracks began to converge on New Jersey, preparations began in earnest. Well before the storm actually hit, evacuation orders had been issued, transportation systems shut down, and emergency relief teams put in place and supplied. These preparations, ultimately traceable to the scientific achievements of the early Bergen meteorologists and their intellectual descendants, doubtless prevented the death toll from climbing much higher than it did.

But Sandy was still an enormous disaster. While it was most disastrous for those in the lowest-lying and hardest-hit coastal areas, the impacts of the infrastructure failures were much broader. The power outages, transportation system closures, gasoline shortages, and other direct and indirect consequences of the storm affected tens of millions of people for days to weeks, and smaller (but still large) numbers for months to years, continuing to the present as I write.

In hindsight, some of the harm could have been reduced by better decisions in the moment, which in turn could have been aided by better communication and understanding of the forecasts. Had NOAA's protocols in 2012 been what they are now, there would almost certainly have been hurricane watches and warnings issued for New York City. Had that happened, perhaps evacuations would have been ordered earlier, and perhaps earlier orders would have convinced some who stayed and suffered death or injury to leave instead. Perhaps nursing homes would have been evacuated ahead of time, while the lights were still on and the transportation systems still fully functioning.

But most of the damage could not have been prevented by any decision made or action taken in the days leading up to the storm. Most of it resulted from decisions made in the years, decades, and even centuries prior. Most was caused, perhaps unconsciously, by decisions to do nothing, or the absence of a decision to do something. These nondecisions were made not in response to scientific predictions, but despite them.

Science allows us to make different kinds of predictions depending on the time frame. In the short term, we can predict the trajectories of specific events. In the long term, we can only make predictions of risk. These are statements about the probability that an event will happen. When a particular type of event has not happened before, predictions that the risk of that event is significant do not, historically, generate the collective will necessary for us to make investments in resiliency. This is true even when the science indicates quite clearly that the event is quite likely to happen eventually, and that the consequences of being unprepared for it will be severe.

Just as the vulnerability of New Orleans was known for decades before Katrina, the vulnerability of New York City and the coastal areas around it was known for decades before Sandy. New York's vulnerability was less than that of New Orleans of course; almost none of the New York metropolitan area is below sea level, and hurricanes come here much less frequently than they come to the Gulf Coast. But perhaps that encouraged even greater disregard for the fragility of the natural environment on which the massive built environment of my hometown sits.

Now the moment has come when we really understand that it can happen here. We understand now that whether we are more prepared next time—not just in what we do in the days leading up to the storm, when our forecasts tell us directly that it is coming, but what we do in the years before to become more resilient—is our responsibility.

As we decide what to do with this new understanding, we have to account for the other kind of forecast that has grown out of the early achievements of modern weather prediction. That is the prediction that the earth is warming due to human emissions of greenhouse gases.

Though chaos theory tells us that we cannot predict the weather more than a couple of weeks ahead, we know that changes in the average of the weather—the climate—can be predicted much farther ahead of time if the causes of the changes are understood. A critically important source of understanding that enables such predictions is our knowledge of the physics of the climate system. This knowledge is encapsulated most explicitly in climate models. These are close relatives of weather prediction models, and inheritors of the tradition started in Bergen. We understand, partly but by no means exclusively because of these models, that carbon dioxide and other pollutants we have added to the atmosphere are warming the climate, and that the warming is exacerbated by natural feedbacks, especially that due to water vapor.

The consequences of global warming are many, and the degrees of certainty with which we know them vary widely. One of the most certain consequences is sea level rise. The rate at which sea level will rise in the coming decades is not certain, but it is almost certain that it will be more rapid than it has been thus far. Several feet of rise are possible, perhaps likely, by the end of the twenty-first century, if not sooner. This alone, without any change in the probability of a storm of any given intensity, will dramatically increase the risk of flood events like Sandy, if serious measures are not put in place to prevent them.

At the same time, it is likely that storms will also respond to the warming climate. The uncertainties here are much greater than those in the rate of global warming itself, but we do have some expectations based on rapidly evolving science of tropical cyclones and their relationship to climate. We expect stronger hurricanes, if perhaps fewer of them worldwide. It is quite possible that the risk of a very strong hurricane will increase. The possible combination of stronger hurricanes with higher sea levels is one of the more distressing risks associated with global warming, though by no means the only one.

That we cannot yet detect any human-induced changes in the observed statistics of tropical cyclones is no cause for complacency. There is such great natural variability from year to year, and decade to decade, that any human influence would have to be very large indeed in order to have been detectable already. Absence of clear statistical evidence for a human influence is not evidence of absence of such an influence.[165] It simply means that we cannot rely solely on historical trends to guide our assessment, because the natural variability is too large and our historical records are too short. We have to use all the scientific tools at our disposal, including computer models that encapsulate our understanding of the physical laws governing the climate.

On the one hand, it may be that changes in tropical cyclones due to climate change are the smaller part of the growing risk of future tropical-cyclone-induced disasters. Increasing coastal development is inarguably a serious cause of increasing risk in the near term, as has been the case in the past, with sea level rise playing an increasingly important role as time goes on. But on the other hand, the considerable uncertainties in our knowledge of the tropical cyclones' response should make us more concerned, not less. When we confront any other serious risk in our lives whose precise level of threat is unknown, does the uncertainty calm us? Or do we take steps to prevent or mitigate the possibility of the worst outcome?

The most important connection of Sandy to global warming may prove to be not in any explicit, demonstrable relationship between the two, but in the storm's role as a signpost. Our difficulties in responding to long-term environmental risks appear to stem from our difficulties in visualizing those risks and understanding them viscerally. When an event is outside our experience, the availability bias makes us blind to the possibility that it may come, even if we know in our rational minds that it well may. We need to experience it in order to prepare better, and farther ahead, for next time.

Narrowly, "next time" means "the next time a major storm threatens a coastal flood in the northeastern United States." The plans laid out in the New York City SIRR report, and the other state and federal documents written after Sandy, could be very effective against that next storm, if acted upon seriously. Elevating and protecting homes, businesses, and infrastructure; finding better ways than plywood to plug subway tunnels; perhaps even convincing more people to move out of the most hazardous areas; and letting some of the former coastal wetlands and barrier islands return to their former states, under the partial ownership of the sea—these things cost money and political will, but they are doable, if we don't forget too quickly what happened when Sandy made landfall.

More broadly, though, "next time" is not just one event, but the foreseeable future, as the climate warms.

For a brief historical moment—as our leaders surveyed the damage and (as all of us in the storm's wake did) tried to process what had happened—it seemed that Sandy might change the discussion on climate here in the United States in a lasting way. Has it done so? Or will our attention now flag, our perception of the reality of global warming's risks dissolve into the inert passivity that is the historical norm?

We respond best to risks for which our prior experience has prepared us. But global warming will bring many changes that are well outside our experience. Sea level rise and changes in storms, though they are the changes most relevant to the story I have told in this book, are just a couple of them. A summer in New York City (or nearly anywhere, for that matter) in 2050 will be unlike anything any of its current living residents have experienced. As sea ice melts, crops fail, and droughts and floods become more severe, the ecology of our planet will be permanently altered. Many of these changes are quite certain, and many are already happening.

But for many of us, the changes will take a long time to have severe and obvious impacts on our lives. The problems will be much more severe when our children and grandchildren grow up, but by then it will be much more difficult to do anything other than adapt to the warming that is now under way. Just like the earlier scientific reports warning what could happen someday if a big storm hit New York City (or New Orleans, or elsewhere), the many authoritative scientific reports that have warned of the risks from climate change still come across to too many as ethereal and vague. They compete poorly with threats that are more immediate (economic crises; diseases and other health risks) or that have active human agents behind them trying to do us harm (terrorism, war).

Most difficult of all, the warming cannot be slowed by easy measures; the gases that are causing it are by-products of virtually everything in our modern economic and energy systems. Besides making all of us complicit, this creates a short-term incentive to deny that the problem exists in the first place. In the United States in particular, that incentive has driven an economically and politically powerful faction to take the position (justifiable at best by ignorance, at worst by willful dishonesty) that it is better to debate the fundamental scientific fact of warming than the choices we should make in response to it. Even among the great number of people who accept the science, the global scale of the actions needed to reduce greenhouse gas emissions significantly makes it difficult to know what to do. But the first thing we need to do is understand in a new way that our natural tendency to ignore remote, unfamiliar risks does not make them go away.

Was Sandy made more probable by global warming? Present science doesn't justify strong answers. But that doesn't matter.

A disaster such as Sandy makes us aware that the climate poses risks. It makes us aware not just intellectually, but at the gut level that we need to be made aware in order to act. How we act in response to global warming should be conditioned on our understanding of the science. But just as the risks aren't limited to those that Sandy made manifest, it isn't necessary that the science pin Sandy on the warming. We already understand the risks well enough to know that action is warranted. We need to learn to take action in advance.

ACKNOWLEDGMENTS

Thanks to:

My editor, Karen Rinaldi, and my agent, Coleen O'Shea, for believing in me, and in the idea of this book, from the start.

Andrew Kruczkiewicz, for helping me to find and organize many of the images in the book, and for dealing with the hassles of getting permission to use them. Suzana Camargo, Robert Hart, Chia-Ying Lee, Linus Magnusson, Philip Orton, Stefan Talke, Michael Ventrice, Yonghui Weng, and Fuqing Zhang for providing scientific graphs and charts.

My scientific colleagues at Columbia, for providing the environment in which I could be the scientist who could write this book, and especially for their insights in so many stimulating conversations on topics related (and un-related) to it. Especially Michela Biasutti, Suzana Camargo, Lorenzo Polvani, and Michael Tippett, with whom I've worked so closely. Suzana in particular; most of my research on tropical cyclones over the last fifteen years has been done in intellectual partnership with her, and that background is largely what gave me the confidence to speak and write so much about Sandy.

All the guest speakers from the impromptu seminar course I organized at Columbia in the spring semester of 2013, for contributing to my understand-ing of Sandy's many dimensions. Many of their insights are in the book in one way or another, probably more than I realize: Sim Aberson, Elizabeth Barnes, Michael Byrne, Suzana Camargo, Tim Creyts, Projjal Dutta, Cheryl Hapke, Solomon Hsiang, Klaus Jacob, Howard Kunreuther, Marit Larson, Ning Lin, Ben Orlove, Philip Orton, Richard Plunz, and Cynthia Rosenzweig.

Kerry Emanuel, who casts a long shadow across the book in several ways, for his encouragement, for his support and wisdom over many years, and for reading the manuscript and providing insightful comments.

Meredith Nettles and Maureen Raymo, for helping me understand sea level. Naomi Oreskes, for insights (and inspiration) on communicating about climate. Cheryl Hapke and William Schwab, for education on barrier islands and beach erosion.

Substantial chunks of the book were written while I was a summer visitor at three European scientific institutions: the Laboratoire de Météorologie Dynamique at the Ecole Normale Supérieure in Paris, the Max Planck Institute for Meteorology in Hamburg, and the Department of Meteorology at Stockholm University. So: thanks to my hosts, Jean-Philippe Duvel, Bjorn Stevens, and Johan Nilsson, for organizing these visits, and to them and their colleagues for scientific discussions, some of which were directly relevant to the book (particularly in Stockholm, where I had the thrill of writing about early meteorological history in exactly the place where some key pieces of it happened).

The staff of the Keringhuis at the Maeslant Barrier, for the tour and for the photos of the barrier. Timothy Dugan, Christopher Way, and Diana Errico-Topolski of the U.S. Army Corps of Engineers for arranging a tour of the Stamford Barrier, and for images of it and of the Providence Barrier; and Marci Montrose for giving me the tour in Stamford.

Wouter-Jan Lippmann and Darrel Kerr, for the ride through Zeeland. Tamara Shapiro and Nathan Kleinman of Occupy, and Jonathan Boulware of the South Street Seaport Museum, for telling me their stories. Roland Madden and Paul Julian, for telling me theirs. To my knowledge, the history of the MJO has not been written for a popular audience before, and it is an honor to have had the opportunity to do it.

The outstanding forecasters and scientists at the U.S. National Hurricane Center, in Miami, Florida, for their forecasts, and also for stimulating and enlightening discussions when I had the opportunity to visit them in January 2013. There isn't much in the book about who they are, but their voices, in all caps, provide the backbone of the first half. There is no more authoritative source on what an Atlantic Hurricane is doing at the time it is doing it than the NHC. I knew that if I were quoting it, I wouldn't have to defend the statements, only interpret them.

My children, Eli and Samuel; my parents, Cynthia and Gerald; and my sister, Melissa, for their love and support.

Most of all, my wife, Marit Larson, for everything.

NOTES

1. "In Memoriam," Calisphere, University of California, March 1976, http://content.cdlib. org/view?docId=hb9k4009c7&chunk.id=div00040&brand=calisphere&doc.view=entire_text; thanks to Paul Julian.
2. J. M. Lewis, M. G. Fearon, and H. E. Kleiforth, "Herbert Riehl: Intrepid and Enigmatic Scholar," *Bulletin of the American Meteorological Society* 93 (2012): 963–85. See quote by R. Bryson, pp. 967–68.
3. The term *radiosonde* refers to weather balloons that communicate their data back to the surface by radio signals. The term *rawinsonde* refers to weather balloons that are tracked by radar from the surface or, more recently, using GPS; the tracking allows the motion of the balloon to be known, which is thus used to infer the winds. Without that tracking, pressure, temperature, and humidity could be measured, but not winds. In recent years, almost all sondes were rawinsondes; the distinction has mostly been lost, and I use the term *radiosonde* generically.
4. Roland Madden, personal communication, Oct. 17, 2013.
5. Paul Julian, personal communication, Oct. 22, 2013.
6. This time series of solar radiation will not be a pure sine curve, because the solar radiation becomes zero at night, rather than negative. Nonetheless, the signal will be dominated by the diurnal frequency, one cycle per day.
7. Between 1953 and 1979, only feminine names were used; before 1953, storms were not named.
8. Universal time (UTC), is the time reference used most commonly in meteorology. It is the same as Greenwich Mean Time (GMT). Local time on the East Coast of the United States is either five hours or four hours behind UTC, depending on the time of year (due to daylight saving time); it was four hours behind UTC in late October 2012.
9. *The Signal and the Noise: Why So Many Predictions Fail—but Some Don't,* by Nate Silver, gives a wonderful overview of statistical prediction in many fields.
10. David B. Stephenson, Heinz Wanner, Stefan Bronnimann, and Jurg Luterbacher, "The History of Scientific Research on the North Atlantic Oscillation," in James. W. Hurrell, Yochanan Kushnir, Geir Ottersen, and Martin Visbeck, eds., *The North Atlantic Oscillation: Climatic Significance and Environmental Impact* (Washington, DC: American

Geophysical Union, 2002), pp. 37–50, citing *The King's Mirror*, a Middle Age text whose author is unknown.

11. D. W. J. Thompson and J. M. Wallace, "The Arctic Oscillation Signature in the Wintertime Geopotential Height and Temperature Fields," *Geophysical Research Letters* 25 (1998): 1297–1300.

12. "Jamaica: Direct Hit from Hurricane Sandy," Associated Press, *New York Times*, Oct. 24, 2012, http://www.nytimes.com/2012/10/25/world/americas/jamaica-direct-hit-from-hurricane-sandy.html?_r=0; and Dr. Jeff Masters, "Hurricane Sandy Hits Jamaica, Dumps Heavy Rains on Haiti," Weather Underground, Oct. 24, 2012, http://www.wunderground.com/blog/JeffMasters/hurricane-sandy-hits-jamaica-dumps-heavy-rains-on-haiti.

13. "Sandy Destroys Homes, Livestock, Crops in St. Thomas and Portland," *Jamaica Observer*, Oct. 26, 2012, http://www.jamaicaobserver.com/news/Sandy-destroys-homes--livestock--crops-in-St--Thomas---Portland.

14. "Haiti Official Says Sandy Destroyed Most Crops in Nation's South; Jamaican PM Reports Price Tag," Associated Press, Oct. 30, 2012, http://www.foxnews.com/world/2012/10/30/official-rains-and-wind-from-sandy-destroyed-70-percent-crops-in-southern-haiti/.

15. Stuart B. Schwartz, "Hurricanes and the Shaping of Circum-Caribbean Societies," *Florida Historical Quarterly* 83, no. 4 (Spring 2005): 381–409.

16. Many sources document Columbus's hurricane. Besides Schwartz, I have consulted Kerry Emanuel's *The Divine Wind: The History and Science of Hurricanes* (New York: Oxford University Press, 2005).

17. Schwartz, "Hurricanes."

18. Erik Larson, *Isaac's Storm: A Man, a Time, and the Deadliest Hurricane in History* (New York: Crown, 1999).

19. A. Habib, Md. Shahidullah, and D. Ahmed, "The Bangladesh Cyclone Preparedness Program. A Vital Component of the Nation's Multi-Hazard Early Warning System," in M. Golnaraghi, ed., *Institutional Partnerships in Multi-Hazard Early Warning Systems* (Berlin: Springer, 2012).

20. J. M. R. Torres and M. A. Puig, "The Tropical Cyclone Early Warning System of Cuba," in Golnaraghi, ed., *Institutional Partnerships*.

21. R. Simpson and R. Riehl, "Mid-Tropospheric Ventilation as a Constraint on Hurricane Development and Maintenance," preprints, Tech. Conf. on Hurricanes, American Meteorological Society, Miami Beach, FL, 1958, D4-1–D4-10; M. Riemer, M. T. Montgomery, and M. E. Nicholls, "A New Paradigm for Intensity Modification of Tropical Cyclones: Thermodynamic Impact of Vertical Wind Shear on the Inflow Layer," *Atmospheric Chemistry and Physics* 10 (2012): 3163–88; B. Tang and K. Emanuel, "Midlevel Ventilation's Constraint on Tropical Cyclone Intensity," *Journal of the Atmospheric Sciences* 67 (2010): 1817–30.

22. Dynamical meteorology involves the application of geophysical fluid dynamics specifically to the atmosphere. Geophysical fluid dynamics is the study of the large-scale motions of both the atmosphere and the ocean. The two fluids have many similarities.

23. James L. Franklin, Michael L. Black, and Krystal Valde, "GPS Dropwindsonde Wind Profiles in Hurricanes and Their Operational Implications," *Weather Forecasting* 18 (2003): 32–44.

24. Until the storm is forecast to die, or actually does die.

25. "Category 2 Sandy Kills 11 in Cuba", MyFoxBoston.com, Oct. 25, updated Oct. 26, http://www.myfoxboston.com/story/19910594/hurricane-sandy-makes-landfall-in-cuba. John Schwartz, "Early Worries That Hurricane Sandy Could Be a 'Perfect Storm'" *New York Times*, Oct. 25, 2012, http://www.nytimes.com/2012/10/26/us/early-worries-that-hurricane-sandy-may-be-a-perfect-storm.html?_r=0.

26. "Hurricane Sandy Has Potential to Be a Super Storm for U.S." CBSNews.com, Oct. 15, 2012, http://www.cbsnews.com/8301-505263_162-57539880/hurricane-sandy-has-potential-to-be-super-storm-for-u.s/.

27. Jennifer Preston, "Tracking Hurricane Sandy up the East Coast," *New York Times*, Oct. 25, 2012, http://thelede.blogs.nytimes.com/2012/10/25/hurricane-sandy-threatens-northeast-and-mid-atlantic/. The article includes a quote by Paul Kocin, a meteorologist for NOAA: "If it actually hits at the kind of intensity, it might across [sic] central to northern New Jersey. That would be a potential worst-case scenario for New York City. That would maximize coastal flooding and winds for New York. The flooding could be of a level that would be rarely ever seen." The article went on to say, "But Mr. Kocin cautioned that forecasts could change and that some computer models were showing that Hurricane Sandy, expected to move this week into the western Atlantic before turning back toward the northern coast, could hit anywhere from the Carolinas to the DelMarva peninsula to the southern coast of New England. What is certain, however, is that the winter weather moving from the Midwest is setting the stage for a dangerous storm."

28. Daniel Kahneman, *Thinking Fast and Slow* (New York: Farrar, Straus and Giroux, 2011).

29. The first weather balloons were launched in France, in 1892. The modern term *radiosonde* reflects this history: the word *sonde* is French for the noun *probe*.

30. This paraphrases Peter Lynch, "From Richardson to Early Numerical Weather Prediction," pp. 3–17, in Leo Donner, Wayne Schubert, and Richard Somerville, eds., *The Development of Atmospheric General Circulation Models: Complexity, Synthesis and Computation* (Cambridge, UK: Cambridge University Press, 2011), pp. 255 and xvi.

31. http://celebrating200years.noaa.gov/foundations/numerical_wx_pred/S1Chart06_full.jpg.

32. E. N. Rappaport and R. H. Simpson, "Impact of Technologies from Two World Wars," in R. Simpson, ed., *Hurricane! Coping with Disaster* (Washington, DC: American Geophysical Union, 2003), pp. 39–61.

33. This "barotropic model" is described in M. DeMaria and J. M. Gross, "Evolution of Prediction Models," pp. 103–26, in Simpson, ed., *Hurricane!*.

34. The suffix *cline* indicates an angle, as in *incline*.

35. S. Fujiwhara, "On the Growth and Decay of Vortical Systems," *Quarterly Journal of the Royal Meteorological Society* 49 (1923): 75–104.

36. K. Jacob, G. Deodatis, J. Atlas, M. Whitcomb, M. Lopeman, O. Markogiannaki, Z. Kennett, A. Morla, R. Leichenko, and P. Vancura, "Transportation," in C. Rosenzweig, W. Solecki, A. DeGaetano, M. O'Grady, S. Hassol, and P. Grabhorn, eds., "Responding to Climate Change in New York State: The ClimAID Integrated Assessment for Effective Climate Change Adaptation," technical report, New York State Energy Research and Development Authority (NYSERDA), Albany, New York, 2011, www.nyserda.ny.gov.

37. "Service Assessment, Hurricane/Post Tropical Cyclone Sandy, October 22–29, 2012," U.S. Department of Commerce, National Oceanic and Atmospheric Administration, National Weather Service, Silver Spring, MD, 2013, http://www.nws.noaa.gov/os/assessments/pdfs/Sandy13.pdf.

38. "Stress" is just the force per unit area. We measure it this way rather than in terms of total force because the sea is not a discrete object. For a given wind, the total force is greater if it is measured over a larger area. Stress can be expressed in the same units as pressure, in that both are forces per area. Pressure is the force per area normal to the surface (i.e., pressing down on the ground, or the sea), while stress refers to a force parallel to the surface.

39. Video of the storm surge caused by Typhoon Haiyan (2013) in the Philippines can be seen at https://www.youtube.com/watch?v=rS0gv4Xbw7w.

40. Each drop in pressure of 1 hectoPascal results in a sea level elevation of approximately 1 centimeter. This number is a simple consequence of the strength of earth's gravity and the density of water. My rough estimate of 2 feet comes from this number, Sandy's central low surface pressure of 946 hPa, and the typical sea level pressure of around 1013 hPa: 1013 – 946 = 67 hPa. If we then estimate that this caused a 67 cm addition to storm surge (right at the center; it would have been less outside the center, where the pressure drop was smaller), that corresponds to about 26 inches, a little more than 2 feet.

41. J. Samenow, "Monster East Coast Storm Next Week or Big Miss?" Capital Weather Gang, *Washington Post*, October 22, 2012, http://www.washingtonpost.com/blogs/capital-weather-gang/post/monster-east-coast-storm-next-week-or-big-miss/2012/10/22/94bc2152-1c72-11e2-9cd5-b55c38388962_blog.html.

42. "Hurricane Sandy: Timeline," FEMA, http://www.fema.gov/hurricane-sandy-timeline.

43. "Bloomberg Takes Blame for Response to Snowstorm," *New York Times*, Dec. 29, 2010, http://www.nytimes.com/2010/12/30/nyregion/30snow.html?_r=0.

44. C. Perrow, "Using Organizations: The Case of FEMA," *Homeland Security Affairs: The Journal of the Naval Postgraduate School Center for Homeland Defense and Security* 1, no. 2 (2005), http://www.hsaj.org/?fullarticle=1.2.4.

45. Satellite image at http://news.yahoo.com/photos/noaa-satellite-image-taken-sunday-oct-28-2012-photo-202528515.html.

46. K. L. Corbosiero and J. Molinari, "The effects of vertical wind shear on the distribution of convection in tropical cyclones," *Monthly Weather Review* 130 (2002): 2110–23.

47. Kerry Burke, Greg B. Smith, and Corky Siemaszko, "Crane Collapse in Midtown Manhattan as Hurricane Sandy Storms into East Coast," *New York Daily News*, Oct.

29, 2012, http://www.nydailynews.com/new-york/crane-collapse-manhattan-article-1.1194790#axzz2ePNSCHKD.

48. Eric W. Sanderson, *Mannahatta: A Natural History of New York City* (New York: Abrams, 2009).

49. K. A. Mooney, A. Cabrera, and M. Naik, "MTA-NYCT's Hurricane Planning and Infrastructure Mitigation," in J. Khinda, ed., *Impact of Sandy's Storm Surge on NY/NJ Infrastructure*, Proceedings of the 2013 Seminar and Exhibition, American Society of Civil Engineers Metropolitan Section Infrastructure Group, pp. 37–49.

50. Robert Sullivan, "Could New York City Subways Survive Another Hurricane?" *New York Times*, Oct. 23, 2013.

51. Patrick G. McHugh, Luciano N. Villani, Zachary Wolf, and Kevin Davis, 2013: "Storm Preparation and Restoration: Con Edison's Response to Hurricane Sandy and Impact on Its Infrastructure," in Jagtar S. Hkinda, ed., *Impact of Sandy's Storm Surge on NY/NJ Infrastructure*, proceedings of the 2013 seminar and exhibition, American Society of Civil Engineers, 2013.

52. According to interviews conducted by Andrew Kruczkiewicz with twenty Rockaways residents for a term paper written for a seminar course on Sandy, Spring 2013, at Columbia University.

53. "Utilities Caused Post-Sandy Fires That Decimated Breezy Point and Turned It into a 'War Zone' 120 Residents Say in Suit," *New York Post*, July 2, 2013, http://nypost.com/2013/07/02/utilities-caused-post-sandy-fires-that-decimated-breezy-point-and-turned-it-into-war-zone-120-residents-say-in-suit/.

54. "Hurricane Sandy Aerial Photos: Staten Island's Oakwood Beach Neighborhood," Silive.com, Staten Island Advance/Bill Lyons, http://photos.silive.com/4499/gallery/hurricane_sandy_aerial_photos_staten_islands_oakwood_beach_neighborhood/index.html#/0.

55. Ian Frazier, "The Toll," *The New Yorker*, Feb. 11, 2013.

56. *A Stronger, More Resilient New York*, Special Initiative for Rebuilding and Resiliency, PlaNYC, New York City, May 2013.

57. "Mapping Hurricane Sandy's Deadly Toll," *New York Times*, Nov. 17, 2012.

58. "48 Hours: Before, during, and after Sandy in Seaside Heights and Cape May," NJ.com, Nov. 4, 2012, http://www.nj.com/news/index.ssf/2012/11/48_hours_before_during_and_aft.html.

59. Memorandum, City of Hoboken, Office of the Business Administrator to Hoboken mayor Dawn Zimmer, http://www.hobokennj.org/docs/sandy/Memo-Hurricane-Sandy-Hoboken-Sequence-of-Events.pdf.

60. Christopher H. Smith (New Jersey politician), "Floor Statement on Sandy Supplemental," Congress of the U.S. House of Representatives, Jan. 2, 2013, http://chrissmith.house.gov/uploadedfiles/floor_remarks_on_sandy_jan_2_2013.pdf. Somewhat lower numbers in "Superstorm Sandy Deaths, Damage, and Magnitude: What We Know One Month Later," Associated Press, Nov. 29, 2012, http://www.huffingtonpost.com/2012/11/29/superstorm-hurricane-sandy-deaths-2012_n_2209217.html.

61. Jennifer Sargent, "Hurricane Sandy Brought Surge in Carbon Monoxide Poisonings," LiveScience, Oct. 28, 2013, http://www.livescience.com/40739-hurricane-sandy-brought-surge-in-carbon-monoxide-poisonings.html.

62. Jennifer Preston, Sheri Fink, and Michael Powell, "Behind a Call That Kept Nursing Home Patients in Storm's Path," *New York Times*, Dec. 2, 2012, http://www.nytimes.com/2012/12/03/nyregion/call-that-kept-nursing-home-patients-in-sandys-path.html?pagewanted=all.

63. Jeanine Ramirez, "NY1 Exclusive: 125 Nursing Home Residents Died within Months of Sandy Evacuation," NY1.com, Oct. 29, 2013, http://www.ny1.com/content/news/sandy_one_year_later/191202/ny1-exclusive--125-nursing-home-residents-died-within-months-of-sandy-evacuation.

64. J. C. Muter and R. M. Garfield, "Estimating Deceased Victim Totals from Hurricane Katrina," American Meteorological Society 2008 Annual Meeting abstract, https://ams.confex.com/ams/88Annual/webprogram/Paper134842.html.

65. "Hurricane Katrina Statistics Fast Facts," CNN, Aug. 23, 2013, http://www.cnn.com/2013/08/23/us/hurricane-katrina-statistics-fast-facts/.

66. Doyle Rice, "Hurricane Sandy, Drought Cost USA $100 Billion," *USA Today*, Jan. 25, 2013, http://www.usatoday.com/story/weather/2013/01/24/global-disaster-report-sandy-drought/1862201/.

67. Raymond Hernandez, "Congress Approves $51 Billion in Aid for Hurricane Victims," *New York Times*, Jan. 28, 2013, http://www.nytimes.com/2013/01/29/nyregion/congress-gives-final-approval-to-hurricane-sandy-aid.html.

68. Mark Trumbull, "Economic Toll of Sandy: Second Only to Katrina?" *Christian Science Monitor,* Nov. 1, 2012, http://www.csmonitor.com/USA/2012/1101/Economic-toll-of-Sandy-Damage-second-only-to-Katrina.

69. "Sandy and the Top 20 Normalized Hurricane Losses," *Roger Pielke Jr.'s Blog*, Oct. 30, 2012, http://rogerpielkejr.blogspot.com/2012/10/sandy-and-top-20-normalized-us.html.

70. "Hurricane Sandy-Nor'easter Situation Report #13," Dec. 3, 2012 (3:00 p.m. EST), U.S. Department of Energy, Office of Electricity Delivery and Energy Reliability, http://www.oe.netl.doe.gov/docs/SitRep13_Sandy-Nor'easter_120312_300PM.pdf.

71. Dana O'Keefe and Alex Kliment, video by shot between Oct. 30 and Nov. 3, 2012, *New York Times*, shown in *Rising Waters*, an exhibit at the Museum of the City of New York, October 29, 2013–April 20, 2014.

72. Winnie Hu, "Mayor Mandates Rationing of Gas to Ease Shortage", *New York Times*, Nov. 8, 2012, http://www.nytimes.com/2012/11/09/nyregion/new-york-city-imposes-gas-rationing-to-ease-shortage.html.

73. Thanks to Pradipta Parhi and Andrew Kruczkiewicz for notes.

74. "Update on City Recovery and Assistance Operations," PR-413-12, Nov. 9, 2012, http://www.nyc.gov/portal/site/nycgov/menuitem.c0935b9a57bb4ef3daf2f1c701c789a0/index.jsp?pageID=mayor_press_release&catID=1194&doc_name=http%3A%2F%2Fwww.nyc.gov%2Fhtml%2Fom%2Fhtml%2F2012b%2Fpr413-12.html&c-c=unused1978&rc=1194&ndi=1.

75. "New York State Homeowners Coverage, Approved Independent Mandatory Hurricane Deductibles," revised Aug. 22 2013, http://www.dfs.ny.gov/consumer/homeown/awindded.pdf.

76. "Sen. Schumer to Insurance Companies: You Will 'Pay Every Dollar for Every Legitimate Claim in the Wake of Sandy,'" *New York Daily News*, Nov. 11, 2012, http://www.nydailynews.com/new-york/schumer-insurance-companies-pay-legitimate-claim-wake-sandy-article-1.1200319#ixzz2wLzy3BwM.

77. "Christie Administration Takes Action to Protect Storm-Impacted New Jersey Homeowners from Higher Insurance Deductibles," State of New Jersey, Office of the Governor, Nov. 2, 2012, http://nj.gov/governor/news/news/552012/approved/20121102e.html.

78. Michael R. Bloomberg, "A Vote for a President to Lead on Climate Change," BloombergView, http://www.bloomberg.com/news/2012-11-01/a-vote-for-a-president-to-lead-on-climate-change.html.

79. Assuming the process doesn't get out of control, leading to a "runaway greenhouse," where the climate becomes *much* hotter. The climate of Venus is believed to be a runaway greenhouse, but this doesn't seem likely to happen in any foreseeable future for Earth.

80. J. Hurley and J. Galewsky, "A Last-Saturation Analysis of ENSO Humidity Variability in the Subtropical Pacific," *Journal of Climate* 23 (2010): 918–31.

81. It is sometimes useful to think of any consequence of human-induced climate change as a result of two components: the global warming itself, as measured by an increase in the global average surface temperature, and the consequence that follows from that, such as sea level rise. To the extent that the system is *linear*, we can imagine that the consequence can be quantified *per degree increase* of global surface temperature. In other words, if one degree global warming causes a given amount of sea level rise, two degrees causes twice as much sea level rise, etc. The problem of understanding sea level rise is thus broken into two simpler problems: the problem of predicting the rate of global warming and the problem of predicting how sea level will respond to each degree of global warming. Multiplying the two tells us the rate of sea level rise. The system is not truly linear, so this kind of thinking has its limitations, but is a useful starting point for many purposes.

82. E.g., R. J. Nicholls and A. Cazenave, "Sea-Level Rise and Its Impact on Coastal Zones," *Science* 328 (2010): 1517–20.

83. Anders Levermann, "The Inevitability of Sea Level Rise," Real Climate, Aug. 15, 2013, http://www.realclimate.org/index.php/archives/2013/08/the-inevitability-of-sea-level-rise/; and A. Levermann, P. U. Clark, B. Marzeion, G. A. Milne, D. Pollard, V. Radic, and A. Robinson, 2013: "The Multimillennial Sea-Level Commitment of Global Warming," *Proceedings of the National Academy of Sciences*, 2013, doi:10.1073/pnas.1219414110.

84. M. O'Leary, P. Hearty, W. Thompson, M. E. Raymo, J. Mitrovica, and J. Webster, "Ice Sheet Collapse Following a Period of Stable Eustatic Sea Level during MIS5e," *Nature Geoscience*, 2013, doi:10.1038/ngeo1890.

85. M. E. Raymo and J. X. Mitrovica, "Collapse of Polar Ice Sheets during the Stage 11 Interglacial," *Nature*, 2012, doi:10.1038/nature10891.

86. For example, see J. X. Mitrovica, N. Gomez, E. Morrow, C. Hay, K. Latychev, and M. E. Tamisiea, "On the Robustness of Predictions of Sea Level Fingerprints," *Geophysical Journal International* 187 (2011): 729–42.

87. A one-hundred-year event—one with a one-hundred-year "return period"—means that there is a 1 percent chance of the event occurring each year; a five-hundred-year return period means there is a 0.2 percent chance each year. The five-hundred-year flood contours in the FEMA flood maps of New York City in place before Sandy were, on average, close to the contours of inundation that actually occurred during Sandy. There had not been a similar coastal flood in the city since at least 1821. Though these facts are not enough to make a precise estimate of the return period, they certainly suggest one greater than one hundred years, perhaps several hundred years. These numbers do not account for sea level rise or any other aspect of climate change, but they are estimates of the probability of a Sandy-like event under the historical climate.

88. J. E. Ellemers, *Studies in Holland Flood Disaster 1953, Volume IV: General Conclusions,* Institut voor Sociaal Onderzoek van het Nederlandse Volk Amsterdam, Committee on Disaster Studies of the National Academy of Sciences—National Research Council, Washington, DC, 1955.

89. I visited the Lake Borgne Surge Barrier to appear in a video made by National Geographic in collaboration with the AXA Research Fund. The video is viewable at https://gallery.axa-research.org/en/webdoc/webdocs/extreme-weather-events.htm.

90. For example, Andrew Higgins, "Lessons for U.S. from a Flood-Prone Land," *New York Times,* Nov. 14, 2012, http://www.nytimes.com/2012/11/15/world/europe/netherlands-sets-model-of-flood-prevention.html?_r=0.

91. Maeslant Barrier museum exhibits.

92. Michael Kimmelman, "Going with the Flow," *New York Times,* Feb. 13, 2013, http://www.nytimes.com/2013/02/17/arts/design/flood-control-in-the-netherlands-now-allows-sea-water-in.html?pagewanted=all&_r=0.

93. Thomas Kaplan, "Cuomo Seeking Home Buyouts in Flood Zones," *New York Times,* Feb. 3, 2013, http://www.nytimes.com/2013/02/04/nyregion/cuomo-seeking-home-buyouts-in-flood-zones.html.

94. Kia Gregory, "Deciding Whether It's Lights Out," *New York Times,* Oct. 25, 2013, http://www.nytimes.com/2013/10/27/nyregion/deciding-whether-its-lights-out.html.

95. McKenzie Funk, "Deciding Where Future Disasters Will Strike," *New York Times,* Nov. 3, 2013, http://www.nytimes.com/2012/11/04/opinion/sunday/deciding-where-future-disasters-will-strike.html.

96. L. Magnusson, J.-R. Bidlot, S. Lang, A. Thorpe, N. Wedi, and M. Yamaguchi, "Evaluation of Medium-Range Forecasts for Hurricane Sandy," *Monthly Weather Review* 142 (2014), doi: http://dx.doi.org/10.1175/MWR-D-13-00228.1.

97. Some of the essential ideas go back to Riehl and Kleinschmidt in the 1950s. Later, nearly contemporaneous with Emanuel, Greg Holland developed a potential intensity theory broadly similar to Emanuel's but different in detail.

98. E.g., G. H. Bryan and R. Rotunno, "Evaluation of an Analytical Model for the Maximum Intensity of Tropical Cyclones," *Journal of the Atmospheric Sciences* 66 (2009): 3042–60; J. Persing and M. T. Montgomery, "Hurricane Superintensity," *Journal of the Atmospheric Sciences* 60 (2003): 2349–71; and R. K. Smith, M. T. Montgomery, and S. Vogl, "A Critique of Emanuel's Hurricane Model and Maximum Potential Intensity Theory," *Quarterly Journal of the Royal Meteorological Society* 134 (2008): 551–61.

99. K. A. Emanuel, 2000: "A Statistical Analysis of Tropical Cyclone Intensity," *Monthly Weather Review* 128 (2000): 1139–52; and A. A. Wing, A. H. Sobel, and S. J. Camargo, "Relationship between the Potential and Actual Intensities of Tropical Cyclones on Interannual Time Scales," *Geophysical Research Letters* 34 (2007), L08810, doi:10.1029/2006GL028581.

100. If they were using a fully coupled climate model with a real ocean, they would have to increase only carbon dioxide; the sea surface temperature increase would result organically from the carbon dioxide increase. Because it's an atmosphere-only model, the global sea surface temperature field is taken from a simulation with another, lower-resolution climate model. Because different climate models produce slightly different patterns of sea surface temperature—all warming, but with slight differences in the overall amount of warming and in how it varies from place to place—for the same carbon dioxide increase, they do the HiRAM calculations several times with sea surface temperatures from different climate models.

101. T. R. Knutson, J. L. McBride, J. Chan, K. Emanuel, G. Holland, C. Landsea, I. Held, J. P. Kossin, A. K. Srivastava, and M. Sugi, "Tropical Cyclones and Climate Change," *Nature Geoscience* 3 (2010): 157–63, doi: 10.1038/ngeo779.

102. E. Palmén, "On the Formation and Structure of the Tropical Hurricane," *Geophysica* 3 (1948): 26–38.

103. J. D. Neelin, C. Chou, and H. Su, 2003, "Tropical Drought Regions in Global Warming and ENSO Teleconnections," *Geophysical Research Letters* 30 (2003), doi:10.1029/2003GL018625.

104. Knutson et al., "Tropical Cyclones."

105. K. A. Emanuel, "Tropical Cyclone Activity Downscaled from NOAA-CIRES Reanalysis, 1908–1958," *Journal of Advances in Modeling Earth Systems* 2 (2010), doi:10.3894/JAMES.2010.2.1. S. J. Camargo, M. K. Tippett, A. H. Sobel, G. A. Vecchi, and M. Zhao, "Testing the Performance of Tropical Cyclone Genesis Indices in Future Climates Using the HiRAM Model," *Journal of Climate*, in review.

106. Three percent is a much, much larger fraction of the air mass than water makes up for the atmosphere as a whole. That is because the saturation-specific humidity is a strong (exponential, in fact) function of temperature. The air at the surface in the tropics is the warmest air in the whole atmosphere. Most of the rest of the atmosphere (both at higher altitudes and higher latitudes) is much colder, and so can hold much, much less water vapor.

107. K. A. Emanuel, "Downscaling CMIP5 climate Models Shows Increased Tropical Cyclone Activity over the 21st Century," *Proceedings from the National Academy of Sciences* 110 (2013), doi/10.1073/pnas.1301293110.

108. For a cartoon dramatization of the NHC advisories during the extreme season of 2005, see http://xkcd.com/1126/.

109. K. A. Emanuel, "Increasing Destructiveness of Tropical Cyclones over the Past 30 Years," *Nature* 436 (2005): 686–88.

110. P. J. Webster, G. J. Holland, J. Curry, and H.-R. Chang, "Changes in Tropical Cyclone Number, Duration, and Intensity in a Warming Environment," *Science* 309 (2005): 1844–46.

111. G. A. Vecchi and B. J. Soden, "Effect of Remote Sea Surface Temperature Change on Tropical Cyclone Potential Intensity," *Nature* 450 (2007): 1066–71.

112. S.-P. Xie, C. Deser, G. A. Vecchi, J. Ma, H. Teng, and A. T. Wittenberg, "Global Warming Pattern Formation: Sea Surface Temperature and Rainfall" *Journal of Climate* 23 (2010); also A. H. Sobel and S. J. Camargo "Projected Future Seasonal Changes in Tropical Summer Climate," *Journal of Climate* 24 (2011): 473–87.

113. J. Leloup and A. Clement, "Why Is There a Minimum in Projected Warming in the Tropical North Atlantic Ocean?" *Geophysical Research Letters* 36 (2009): L14802, doi:10.1029/2009GL038609.

114. K. A. Emanuel and A. H. Sobel, "Response of Tropical Sea Surface Temperature, Precipitation, and Tropical Cyclone-Related Variables to Changes in Global and Local Forcing," *Journal of Advances in Modeling Earth Systems* 5 (2013): 1–12.

115. G. A. Vecchi and B. J. Soden, "Increased Tropical Atlantic Wind Shear in Model Projections of Global Warming," *Geophysical Research Letters* 34 (2007): L08702, doi:10.1029/2006GL028905.

116. P. A. DiNezio, A. Clement, G. Vecchi, B. Soden, B. Kirtman, and S.-K. Lee, "Climate Response of the Equatorial Pacific to Global Warming," *Journal of Climate* 22 (2009): 4873–92.

117. J. A. Knaff, S. P. Longmore, and D. A. Molenar, "An Objective Satellite-Based Tropical Cyclone Size Climatology," *Journal of Climate* 27 (2013): 455–76, http://dx.doi.org/10.1175/JCLI-D-13-00096.1.

118. J. A. Francis and S. J. Vavrus, "Evidence Linking Arctic Amplification to Extreme Weather in Mid-Latitudes," *Geophysical Research Letters* 39 (2012): L06801, doi:10.1029/2012GL051000; also J. Liu, J. A. Curry, H. Wang, M. Song, and R. M. Horton, 2012: "Impact of Declining Arctic Sea Ice on Winter Snowfall," *Proceedings of the National Academy of Sciences* 109 (2012): 4074–79.

119. E. A. Barnes, L. M. Polvani, and A. H. Sobel, 2013: "Model Projections of Atmospheric Steering of Sandy-Like Superstorms," *Proceedings of the National Academy of Sciences* doi:10.1073/pnas.1308732110.

120. C. Wang, H. Liu, S. K. Lee, and R. Atlas, "Impact of the Atlantic Warm Pool on United States Landfalling Hurricanes," *Geophysical Research Letters* 38 (2011). L19702, doi:10.1029/2011GL049265.

121. IPCC, "Summary for Policymakers," in T. F. Stocker, D. Qin, G.-K. Plattner, M. Tignor, S. K. Allen, J. Boschung, A. Nauels, Y. Xia, V. Bex, and P. M. Midgley, eds., *Climate Change 2013: The Physical Science Basis. Contribution of Working Group I to the Fifth Assess-*

ment Report of the Intergovernmental Panel on Climate Change (Cambridge, UK, and New York: Cambridge University Press, 2013), http://www.climatechange2013.org/images/report/WG1AR5_SPM_FINAL.pdf.

122. M. E. Mann and K. A. Emanuel, "Atlantic Hurricane Trends Linked to Climate Change," *Eos* 87 (2006): 233, 238, 241.

123. M. E. Mann and K. A. Emanuel, Atlantic hurricane trends linked to climate change. EOS, Transactions American Geophysical Union, 87, DOI: 10.1029/2006EO240001 (2006).

124. J. Bjerknes, "Atmospheric Teleconnections from the Equatorial Pacific," *Monthly Weather Review* 97 (1969): 163–72.

125. R. Seager and G. A. Vecchi, "Greenhouse Warming and the 21st Century Hydroclimate of Southwestern North America," *Proceedings of the National Academy of Sciences* 107 (2010): 21277–82.

126. This and other facts about the 1938 storm from Lourdes B. Avilés, *Taken by Storm 1938: A Social and Meteorological History of the Great New England Hurricane* (Washington, DC: American Meteorological Society, 2013); and William Elliott Minsinger, comp. and ed., *The 1938 Hurricane: An Historical and Pictorial Summary* (East Milton, MA: Blue Hill Observatory, 1988).

127. Charles F. Brooks, writing in Minsinger, ed., *The 1938 Hurricane,* cites the American Red Cross death toll as 488, the WPA survey as 682. The other statistics in this paragraph are taken from the same account.

128. John Rather, "Dreading a Replay of the 1938 Hurricane," *New York Times,* Aug. 28, 2005. Charles F. Brooks, in Minsinger, ed., *The 1938 Hurricane,* gives the total as four hundred million dollars rather than three hundred million.

129. Minsinger claims 1938 was the most expensive to date, fifty million dollars more expensive than the 1906 San Francisco earthquake, but it appears to me that his numbers are not adjusted for inflation. Other sources rank the 1900 Galveston hurricane and a few other earlier disasters as more costly than 1938.

130. This and much of the next few paragraphs are from Rutherford H. Platt, H. Crane Miller, Timothy Beatley, Jennifer Melville, and Brenda G. Mathenia, *Coastal Erosion: Has Retreat Sounded?* Program on Environment and Behavior Monograph No. 53, Institute of Behavioral Science, University of Colorado, 1992; and Orrin H. Pilkey and Katharine L. Dixon, *The Corps and the Shore* (Washington, DC: Island Press, 1996).

131. Beach Erosion Board, *Shore Protection Planning and Design,* Office of the Chief of Engineers, Department of the Army, Corps of Engineers, Report No. 4, 1954, Babel.hathitrust.org, original from University of Michigan, digitized by Google.

132. E.g., Larry Savadove and Margaret Thomas Buchholz, *Great Storms of the Jersey Shore* (West Creek, NJ: Down the Shore Publishing, 1997).

133. Coastal Planning and Engineering, Inc., "Coastal Protection Study, City of Long Beach, NY: Oceanside Shore Protection Plan," 2009, http://www.longbeachny.gov/vertical/sites/%7BC3C1054A-3D3A-41B3-8896-814D00B86D2A%7D/uploads/%7B413002EC-E9A4-45AA-8AB7-D193D9294877%7D.PDF.

134. Mireya Navarro and Rachel Nuwer, "Resisted for Blocking the View, Dunes Prove They Blunt Storms," *New York Times*, Dec. 3, 2012.

135. "Long Beach after Sandy," TimesVideo by Vijai Singh and Michael Winerip, *New York Times*, Nov. 2, 2012, http://www.nytimes.com/video/nyregion/100000001880340/long-beach-after-sandy.html.

136. Jim Axelrod, "Dune Fortification Still Controversial Despite Sandy Destruction," Sept. 5, 2013, CBS News, http://www.cbsnews.com/8301-18563_162-57601607/.

137. John Seabrook, "The Beach Builders: Can the Jersey Shore Be Saved?" *New Yorker*, July 22, 2013.

138. Vivien Gornitz, *Rising Seas: Past, Present, Future* (New York: Columbia University Press, 2013).

139. C. J. Hapke, O. Brenner, R. Hehre, and B. J. Reynolds, "Coastal Change from Hurricane Sandy and the 2012–13 Winter Storm Season: Fire Island, New York," Open-File Report 2013–1231, U.S. Geological Survey, U.S. Department of the Interior, http://pubs.usgs.gov/of/2013/1231/pdf/ofr2013-1231.pdf.

140. Cheryl Hapke, personal communication.

141. R. A. Pielke Jr., J. Gratz, C. W. Landsea, D. Collins, M. A. Saunders, and R. Musulin, "Normalized Hurricane Damages in the United States: 1900–2005," *Natural Hazards Review* 9, no. 1 (2008): 29–42.

142. D. King, J. Davidson, and L. Anderson-Berry, "Disaster Mitigation and Societal Impacts," in J. C. L. Chan and J. D. Kepert, eds., *Global Perspectives on Tropical Cyclones* (Hackensack, NJ: World Scientific, 2010), pp. 409–36.

143. Nice discussion in Erik Larson's *Isaac's Storm*.

144. Emanuel, *The Divine Wind*, pp. 106–7.

145. K. T. Jackson, ed., *The Encyclopedia of New York City* (New Haven, CT: Yale University Press, 1995).

146. "Katrina Forecasters Were Remarkably Accurate" NBC News, Sept. 19, 2005, http://www.nbcnews.com/id/9369041/ns/us_news-katrina_the_long_road_back/t/katrina-forecasters-were-remarkably-accurate/#.Uk13eWRASEc.

147. John Rather, "Dreading a Replay of the 1938 Hurricane", *New York Times*, Aug. 28, 2005.

148. "Interim Technical Data Report, Metro New York Hurricane Transportation Study, November 1995," U.S. Army Corps of Engineers, FEMA, National Weather Service, NY/NJ/CT State Emergency Management, 73 pp.

149. The Saffir-Simpson scale was revised in 2009 so that storms would be classified only by their maximum sustained surface winds.

150. Rosenzweig, C., and W. Solecki, eds., "Climate Change and a Global City: The Potential Consequences of Climate Variability and Change Metro East Coast. Report for the U.S. Global Change Research Program, National Assessment of the Potential Consequences of Climate Variability and Change for the United States," Columbia Earth Institute, Columbia University, 2001.

151. K. Jacob, G. Deodatis, J. Atlas, M. Whitcomb, M. Lopeman, O. Markogiannaki, Z. Kennett, A. Morla, R. Leichenko, and P. Vancura, "Transportation," in C. Rosenzweig, W.

Solecki, A. DeGaetano, M. O'Grady, S. Hassol, and P. Grabhorn, eds. "Responding to Climate Change in New York State: The ClimAID Integrated Assessment for Effective Climate Change Adaptation," technical report, New York State Energy Research and Development Authority (NYSERDA), Albany, New York, 2011, www.nyserda.ny.gov.

152. Klaus Jacob, personal communication. The FEMA maps show contours of inundation for a one-hundred-, two-hundred-, and five-hundred-year flood. Sandy, like any real event, did not produce inundation exactly identical to any of those contours, but comes closest to the five-hundred-year contour, on average.

153. Howard C. Kunreuther and Mark V. Pauly, *Insurance and Behavioral Economics: Improving Decisions in the Most Misunderstood Industry* (Cambridge, UK: Cambridge University Press, 2013).

154. Daniel Kahneman, *Thinking, Fast and Slow* (New York: Farrar, Straus and Giroux, 2011). The original reference on availability appears to be A. Tversky and D. Kahneman, "Judgment under Uncertainty: Heuristics and Biases," *Science* 185 (1974): 1124–31.

155. The paper that Tim Hall and I wrote on the left turn has been misinterpreted to mean this.

156. Engineers, on the other hand, use the word *mitigation* to describe what the climate community calls *adaptation*.

157. What controls the total magnitude of the climate change we have locked in, and will eventually experience, is essentially the total amount of carbon we have put into the atmosphere. It follows that a future reduction in emissions doesn't do nearly as much as a reduction now.

158. "New York City Mayor's Carbon Challenge Progress Report, April 2013," http://www.nyc.gov/html/gbee/downloads/pdf/mayors_carbon_challenge_progress_report.pdf.

159. There are sections on Coastal Protection, Buildings, Insurance, Utilities, Liquid Fuels, Healthcare, Telecommunications, Transportation, Parks, Water and Wastewater, and Other Critical Networks.

160. Anne Siders, "New FEMA Flood Maps for New York Do Not Consider Sea Level Rise," Climate Law Blog, Feb. 14, 2013, http://blogs.law.columbia.edu/climatechange/2013/02/14/new-fema-flood-maps-for-new-york-do-not-consider-sea-level-rise/; and Katherine Bagley, "Climate Change Impacts Absent from FEMA's Redrawn NYC Flood Maps," InsideClimate News, Feb. 6, 2013, http://insideclimatenews.org/news/20130204/climate-change-global-warming-flood-zone-hurricane-sandy-new-york-city-fema-federal-maps-revised-sea-level-rise.

161. "Rail on Guard: Metro-North Prepares for Threat of Rising Sea Levels," *Poughkeepsie Journal*, April 22, 2014, http://www.poughkeepsiejournal.com/story/tech/science/environment/2014/04/21/rail-on-guard-metro-north-prepares-for-threat-of-rising-sea-levels/7987279/.

162. NYS 2100 Commission, "Recommendations to Improve the Strength and Resilience of the Empire State's Infrastructure," http://www.governor.ny.gov/assets/documents/NYS2100.pdf.

163. Thomas Kaplan, "Cuomo, Joined by Biden, Details Disaster Aid Plans," *New York Times*,

Jan. 7, 2014, http://www.nytimes.com/2014/01/08/nyregion/cuomo-joined-by-biden-details-disaster-aid-plans.html?_r=0.

164. Joshua Dawsey, "Cuomo Lays out Plan for Sandy Recovery Money," *Wall Street Journal*, Jan. 7, 2014, http://online.wsj.com/news/articles/SB1000142405270230461740457930703 3953274754.

165. For example, see "The Most Common Fallacy in Discussing Extreme Weather Events," RealClimate, March 25, 2014, http://www.realclimate.org/index.php/archives/2014/03/the-most-common-fallacy-in-discussing-extreme-weather-events/.

SELECTED BIBLIOGRAPHY

Where Weather Forecasts Came From

Bjerknes, V. "The Problem of Weather Prediction, as Seen from the Standpoints of Mechanics and Physics." Trans. Allen Greenberg, NOS. http://www.history.noaa.gov/stories_tales/bjerknes.html.

Friedman, R. M. *Appropriating the Weather: Vilhelm Bjerknes and the Construction of a Modern Meteorology.* Ithaca, NY: Cornell University Press, 1989, 251 pp.

Lynch, P. *The Emergence of Numerical Weather Prediction: Richardson's Dream.* Cambridge, UK: Cambridge University Press, 2006, 279 pp.

Phillips, N. A. "Carl-Gustaf Rossby: His Times, Personality, and Actions." *Bulletin of the American Meteorological Society* 79 (1998): 1097–112.

Platzman, George. "The Atmosphere—A Challenge: Charney's Recollections." Interview of Jule Charney. In R. S. Lindzen, E. N. Lorenz, and G. Platzman, eds. *The Atmosphere—A Challenge: The Science of Jule Gregory Charney.* Boston, MA: American Meteorological Society, 1990, pp. 11–82.

Richardson, L. F. *Weather Prediction by Numerical Process.* Cambridge, UK: Cambridge University Press, 1922, 236 pp.

Thorpe, A. J., H. Volkert, and M. J. Ziemanski. "The Bjerknes Circulation Theorem: A Historical Perspective." *Bulletin of the American Meteorological Society* 84 (2003): 471–80.

Saturday, October 27

"Service Assessment: Hurricane/Post-Tropical Cyclone Sandy, October 22–20, 2012." U.S. Department of Commerce. National Oceanic and Atmospheric Administration. National Weather Service, Silver Spring, MD, May 2013.

Zeeland

Deltapark Neeltje Jans and Florad Marketing Group. *The Delta Project: Preserving the Environment and Securing Zeeland against Flooding,* 32 pp. promotional booklet from Deltapark Neeltje Jans, the museum and entertainment complex on Neeltje Jans, the artificial island in the middle of the Oosterschelde barrier, 2012.

Gerritsen, Herman. "What Happened in 1953? The Big Flood in the Netherlands in Retrospect." *Philosophical Transactions of the Royal Society A 363* (2005): 1271–91.

INDEX